The Institute of Biology's Studies in Biology no. 138

The Biology of Parasitic Protozoa

John R. Baker
B.Sc., Ph.D., D.Sc., M.A.,
F.I. Biol.
Senior Officer,
Culture Centre of Algae and Protozoa,
Cambridge

Edward Arnold

First published 1982
by Edward Arnold (Publishers) Limited
41 Bedford Square, London WC1 3DQ

British Library Cataloguing in Publication Data

Baker, John R.
The biology of parasitic protozoa –
(The Institute of Biology's studies in
biology, ISSN 0537-9024; no. 138)
1. Parasitology
I. Title II. Series

ISBN 0–7131–2837–2

Photoset and printed by Photobooks (Bristol) Ltd.

General Preface to the Series

Because it is no longer possible for one textbook to cover the whole field of biology while remaining sufficiently up to date, the Institute of Biology proposed this series so that teachers and students can learn about significant developments. The enthusiastic acceptance of 'Studies in Biology' shows that the books are providing authoritative views of biological topics.

The features of the series include the attention given to methods, the selected list of books for further reading and, wherever possible, suggestions for practical work.

Readers' comments will be welcomed by the Education Officer of the Institute.

1982

Institute of Biology
41 Queen's Gate
London SW7 5HU

Preface

Parasitology is a branch of ecology (from Greek *oikos*, a house and *logia*, branch of knowledge). Concentrating on either parasite or host results in a one-sided view of what is really a dynamic tension. Parasites and their hosts live in a more-or-less stable, though precarious, equilibrium. A small tilt in one direction can lead to the death of the host; a small tilt the other way can result in the death of the parasite. Either consequence damages the parasite – and perhaps the host also. A host supporting a limited number of uninvited guests with little or no untoward effect, in a balanced relationship, may benefit from a state of active immunity ("premunition") which prevents its inundation by re-infection from the environment. However, the biological situation is vastly more complex, and an understanding of it, in spite of about 100 years of classical parasitology, is only just beginning. This book is an attempted introduction to a study of the inter-relationships between parasitic protozoa and their hosts.

Cambridge, 1982

J.

Contents

1 Introduction

1.1 The concept of parasitism

Parasitism has been variously defined and subdivided. A simple scheme is to divide it into three categories. Firstly, *commensalism*, in which one organism (the parasite) uses another (the host) essentially for shelter only, though it often benefits by using some of the host's own food supply. The host does not suffer significantly from this type of association. Common examples include many of the protozoan inhabitants of the intestines of vertebrates, including humans. A few hundred amoebae of, say, the genus *Endolimax* living in the intestine discommode us only marginally by consuming an infinitesimal amount of our daily food intake. Secondly, there is the association called *symbiosis*. In this, both partners benefit: one, or both, may be unable to survive without the other. Common examples are the myriad ciliates (Fig. 1–1) inhabiting the rumen of herbivorous mammals like sheep and cattle; the ciliates depend absolutely on the vertebrate host for shelter and food supply, while the vertebrate benefits from protein and other nutrients synthesized by the protozoa from the otherwise

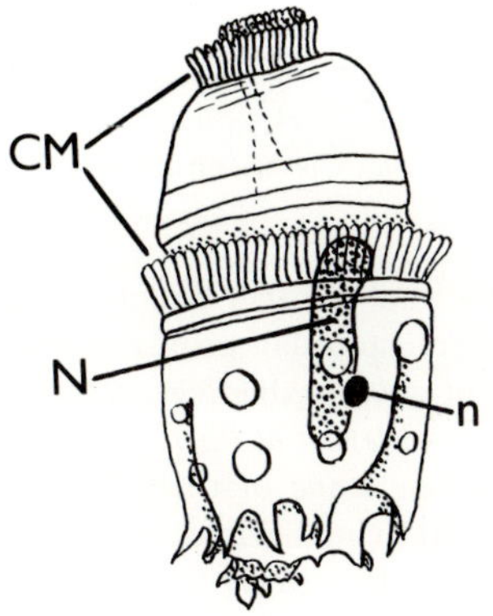

Fig. 1–1 Symbiotic ciliate *(Ophryoscolex caudatus)* from bovine rumen, × 250. CM. ciliary membranelles (fused cilia); N, macronucleus; n, micronucleus. (From BAKER, 1973, after Levine, 1961, *Protozoan Parasites of Domestic Animals and of Man*. Burgess Publishing Co., Minneapolis.)

indigestible portions of their food. The ruminants can, however, survive without the protozoa, though they do so less well. Another, more extreme example is provided by certain flagellates (Hypermastigida) living in the guts of wood-eating insects – cockroaches (Orthoptera) and termites (Isoptera). The flagellates (Fig. 1–2) cannot survive without the shelter and food supplied by the insects, while the latter are equally dependent on their gut fauna since they do not secrete a cellulolytic enzyme and therefore cannot digest their sole food – wood. If the flagellates are killed by warming the host, the latter starves to death unless it can be re-infected with protozoa. The third kind of association is *tissue parasitism*. Some writers restrict the term 'parasitism' to this kind of association alone. Tissue parasitism is an association between two animals of different species in which one (the parasite) lives, temporarily or permanently, in the body of another (the host) and feeds on the latter's tissues. Whereas nearly all pathogenic (i.e. disease-producing) parasitic protozoa are tissue parasites, not

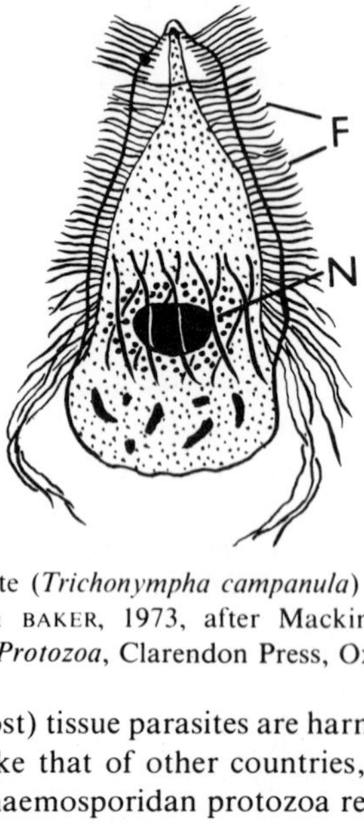

Fig. 1–2 Symbiotic flagellate (*Trichonympha campanula*) from termite's gut, × 200. F, flagella; N, nucleus. (From BAKER, 1973, after Mackinnon and Hawes, 1961, *An Introduction to the Study of Protozoa*, Clarendon Press, Oxford.)

all (probably not even most) tissue parasites are harmful. A high proportion of the British wild fauna, like that of other countries, harbours tissue parasites (often trypanosomes, or haemosporidan protozoa resembling those that cause malaria in humans), without obvious ill effect. A few tissue parasites may even be beneficial (e.g. *Trypanosoma lewisi* in *Rattus* spp., Fig. 1–3). Some parasites may be tissue parasites at one stage of their life cycle, and commensals at another. The balance between living harmlessly in the host and causing disease of the latter may be very delicate – easily tipped one way or the other by, for example, the host's general health or nutritional state. Parasites often have two

Fig. 1–3 *Trypanosoma lewisi*, × 2000 (see Fig. 1–7 for labelling of organelles). (From BAKER, 1973.)

(or more, though not amongst the protozoa) hosts of different species during their life cycle; often one is a vertebrate and the other, called the vector, an invertebrate.

A recent book in this series (LYONS, 1978) has well summarized the advantages and disadvantages of parasitism for helminths; they are fundamentally the same for protozoan parasites. In this book I hope to show how the protozoa have used the advantages and overcome the disadvantages.

1.2 The scope of this book

This short book cannot be comprehensive: it is intentionally eclectic, illustrating certain aspects of the life of parasitic protozoa by referring to examples. The examples will be mainly among parasites of economic importance, if only because such parasites have been more fully studied. Other, less 'important' parasites will be included where they are the best available examples. More complete treatment of many of the aspects mentioned in this book can be found in the books and papers suggested for further reading or the relevant scientific journals, p. 56. Further information about the more economically important parasitic protozoa has been tabulated in the Appendix.

1.3 An outline of systematics

Lengthy discussion of systematics (classification) would be out of place here, and can be found in some of the general reference books listed. But it is convenient to have a brief classification of the parasites referred to in the book. The Protozoa are nowadays usually treated as a subkingdom, with the major subdivisions constituting phyla. If one feels that the absence of photosynthetic pigments from the protozoa is of major taxonomic significance, then the subkingdom can be regarded as part of the kingdom Animalia. If, however, one feels that it is the unicellularity of the protozoa which is significant, remembering that presence or absence of chlorophyll can be a very capricious feature even within a genus (e.g. *Euglena*, normally a green alga but with some colourless species or strains, Fig. 1–4), then it seems more satisfactory to unite the protozoa and unicellular algae in the kingdom Protista. This concept, however, if strictly adhered to, would result in excluding the multicellular myxosporea (Fig. 1–5) and the colonial algae, yet another rather arbitrary distinction. Perhaps the best available solution is to regard the organizational simplicity of non-differentiation into discrete tissues as the dividing line between 'lower' and 'higher' eukaryotes. (Eukaryotic organisms possess mitochondria, nuclear membranes, mitotic spindles, endoplastic reticulum, Golgi apparatus, and '9 + 2' flagella or cilia, and are differentiated from the prokaryotic Monera – bacteria and blue-green algae – which lack all these structures, and also plasmids, and differ biochemically also; Fig. 1–6.) The non-tissue level, including protozoa and algae (and, in some schemes, fungi), can then be accommodated as subkingdoms of the kingdom Protista (or, perhaps more correctly, though less euphoniously, Protoctista).

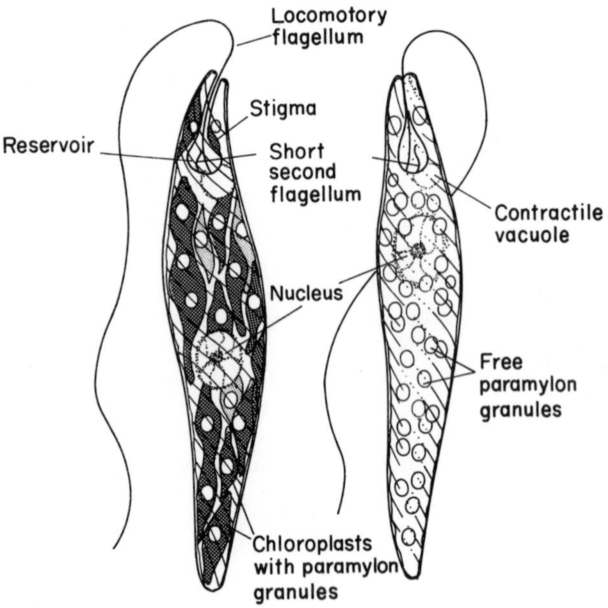

Fig. 1–4 Euglenoid flagellates with (left) and without (right) chloroplasts; note similarity in all other respects. (From Vickerman, K. and Cox, F. E. G., 1967, *The Protozoa*, Murray, London.)

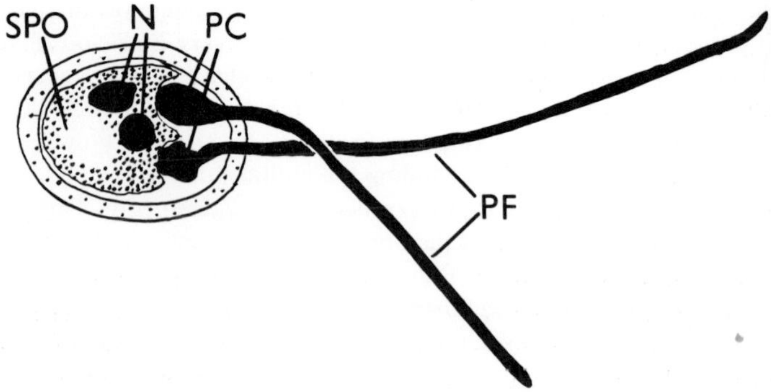

Fig. 1–5 Multicellular spore of *Myxobolus* sp., ×2000. N, nuclei of binucleate cell forming sporoplasm (SPO); PC, unicellular polar capsules, each producing a polar filament (PF); the spore wall is formed from 2 other cells. (From BAKER 1973.)

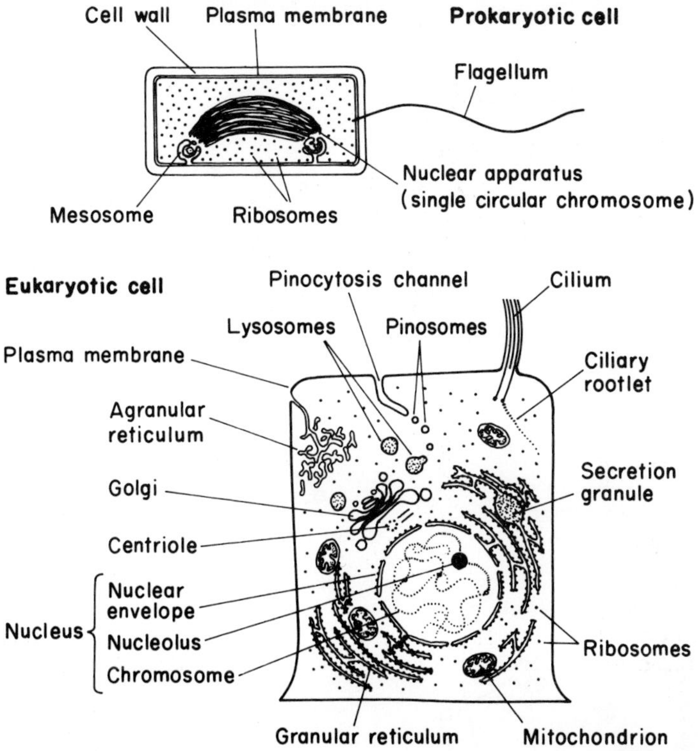

Fig. 1–6 Diagrams comparing structure of prokaryotic and eukaryotic cells; the former comprise bacteria and blue-green algae, the latter all other algae and higher plants, protozoa and multicellular animals: protozoa contain some or all of the structures shown – and others, including microfilaments and microtubules. (From Vickerman, K. and Cox, F.E.G., 1967, *The Protozoa*, Murray, London.)

Precise definition and discrimination is obviously impossible and probably unimportant; clearly there exists a continuum of life in which only division into more-or-less obvious levels of organization (virus, prokaryote, eukaryote) is strictly reasonable (even *Homo sapiens* is unicellular at one stage of the life cycle – the undivided egg). None-the-less, rather less comprehensive divisions, though increasingly arbitrary, are convenient and in general can be equated to more-or-less different levels or types of organization. Within the protozoa, these correspond to the phyla and lesser categories. All the protozoan phyla contain parasitic genera, the Apicomplexa, Microspora and Myxozoa consisting entirely of such forms. There are two major subdivisions of the Sarcomastigophora, depending mainly on the organism's method of locomotion – Mastigophora

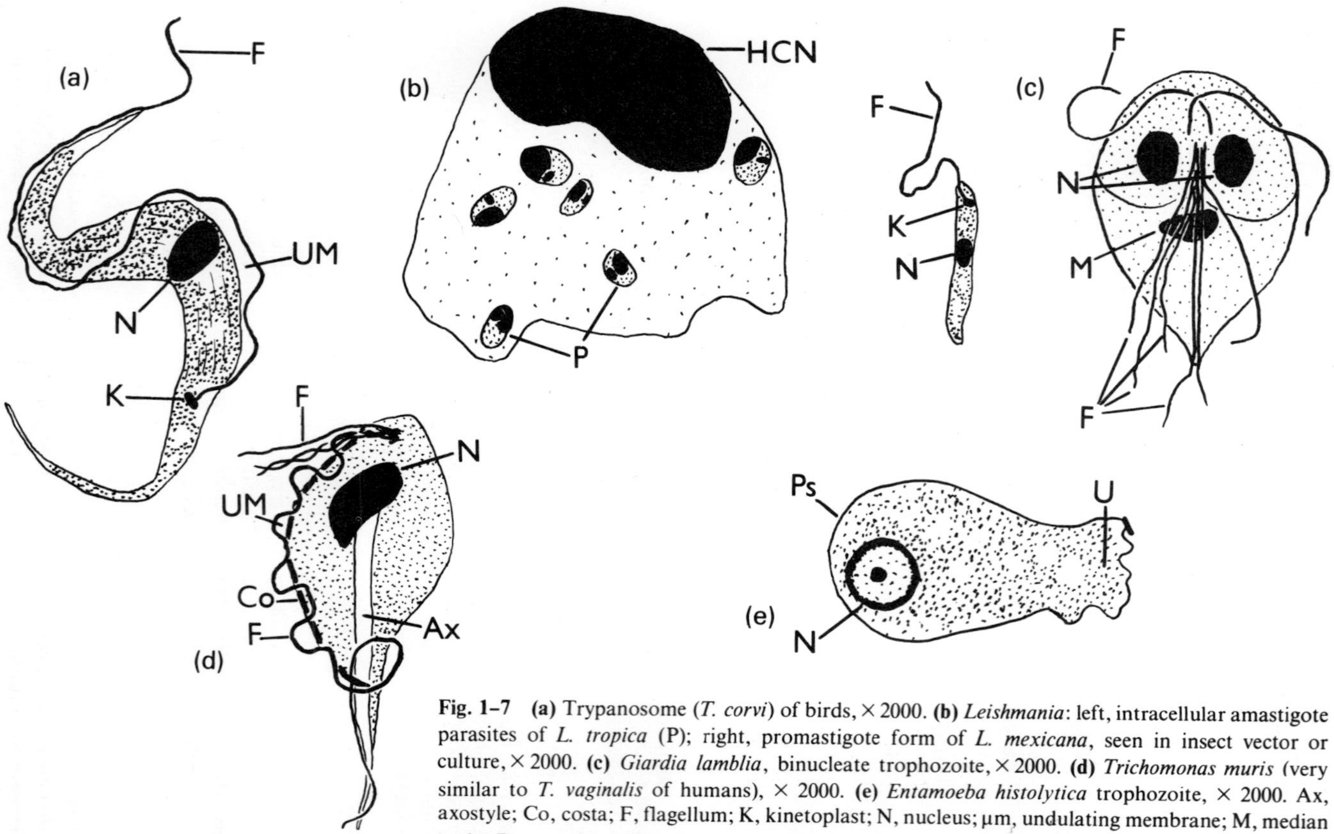

Fig. 1–7 **(a)** Trypanosome (*T. corvi*) of birds, × 2000. **(b)** *Leishmania*: left, intracellular amastigote parasites of *L. tropica* (P); right, promastigote form of *L. mexicana*, seen in insect vector or culture, × 2000. **(c)** *Giardia lamblia*, binucleate trophozoite, × 2000. **(d)** *Trichomonas muris* (very similar to *T. vaginalis* of humans), × 2000. **(e)** *Entamoeba histolytica* trophozoite, × 2000. Ax, axostyle; Co, costa; F, flagellum; K, kinetoplast; N, nucleus; μm, undulating membrane; M, median body; Ps, pseudopodium; U, unroid or 'tail'. (From BAKER, 1973.)

(flagellates) and Sarcodina (amoebae): the former includes important parasites of blood and other tissues (the trypanosomes and leishmanias, Figs. 1–7a and b) and others of the gut (*Giardia*, Fig. 1–7c) or urinogenital tract (*Trichomonas*, Fig. 1–7d), while the latter includes the notorious dysentery amoeba (*Entamoeba histolytica*, Fig. 1–7e). The Apicomplexa comprise three main groups – Gregarinia (Fig. 1–8a) (parasites mainly of invertebrates), Coccidia (including malaria parasites, Fig. 1–8b, and many other important forms) and Piroplasmia (Fig. 1–8c). The Ciliophora contains a vast number of species, some very highly organized, but relatively few parasitic. One (*Balantidium coli*, Fig. 1–9) can cause dysentery in man, and another interesting group lives in the rumen of ruminant, herbivorous mammals (Fig. 1–1); they are commensals which considerably benefit their hosts (see Table 1).

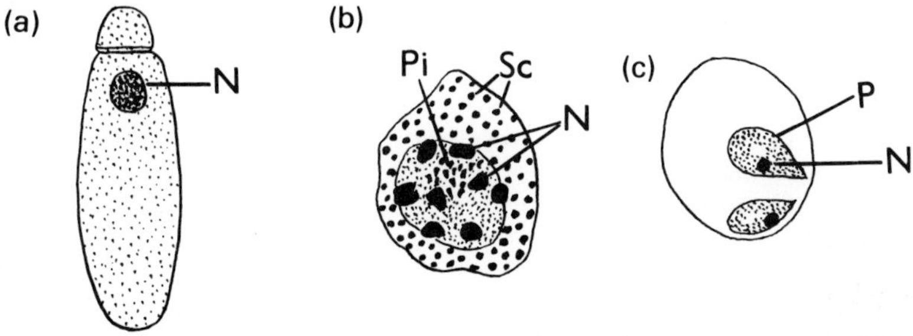

Fig. 1–8 (**a**) *Gregarina ovata*, a gregarine from the gut of an earwig, × 2000. N, nucleus. (**b**) Schizont (division stage) of *Plasmodium vivax* within human red blood cell, × 2000. N, nuclei; Pi, malarial pigment (undigested residue of host cell's haemoglobin); Sc, Schüffner's dots (lesions on surface of infected red cell). (**c**) Two piroplasms (*Babesia canis*) within red blood cell of dog, × 2000. N, nucleus of piroplasm, P. (From BAKER, 1973.)

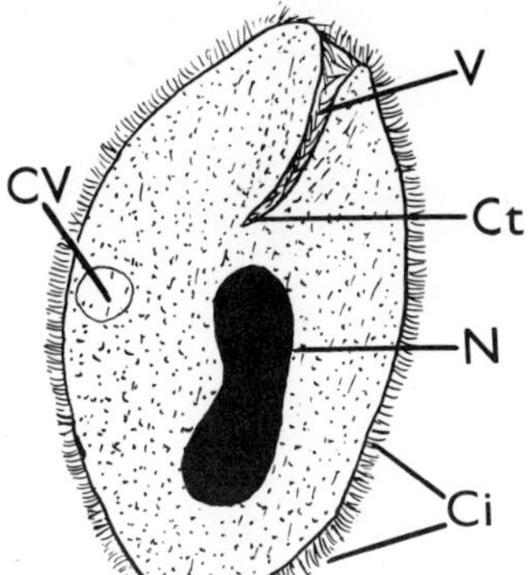

Fig. 1–9 *Balantidium coli* trophozoite, × 650. Ci, cilia; Ct, cytostome ('mouth') at base of vestibule (V); CV, contractile vacuole; N, macronucleus. (From BAKER, 1973.)

Table 1. Outline classification of some parasitic protozoa.

SUBKINGDOM PROTOZOA
- **Phylum Sarcomastigophora**
  - Subphylum Mastigophora
    - Class Zoomastigophorea
      - ORDER Kinetoplastida: *Leishmania, Trypanosoma*, etc.
      - ORDER Diplomonadida: *Giardia*, etc.
      - ORDER Trichomonadida: *Trichomonas, Histomonas*, etc.
      - ORDER Hypermastigida: *Lophomonas, Trichonympha*
  - Subphylum Opalinata
    - Class Opalinatea
      - ORDER Opalinida: *Opalina*
  - Subphylum Sarcodina
    - Superclass Rhizopoda
      - Class Lobosea
        - ORDER Amoebida: *Entamoeba, Acanthamoeba*, etc.
        - ORDER Schizopyrenida: *Naegleria*
- **Phylum Apicomplexa**
  - Class Sporozoea
    - Subclass Coccidia
      - ORDER Eucoccidiida
        - SUBORDER Eimeriina: *Eimeria, Isospora, Toxoplasma, Sarcocystis*, etc.
        - SUBORDER Haemosporina: *Plasmodium*, etc.
    - Subclass Piroplasmia
      - ORDER Piroplasmida: *Babesia, Theileria*, etc.
- **Phylum Microspora**
  - Class Microsporea
    - ORDER Microsporida: *Nosema*, etc.
- **Phylum Myxozoa**
  - Class Myxosporea
    - ORDER Bivalvulida: *Myxobolus*, etc.
- **Phylum Ciliophora**
  - Class Kinetofragminophorea
    - Subclass Vestibuliferia
      - ORDER Entodiniomorphida: *Entodinium*, etc. ('rumen ciliates')

2 Finding and Infecting the Host

This vital problem has been solved in several ways. The least specialized are the contaminative methods adopted by parasitic protozoa having only a single host species (*monoxenous* parasites), though these, involving a sojourn in the external environment, are the most hazardous solutions. Other parasites, with hosts of two different species (*dixenous* parasites), have developed more sophisticated means of transport from one host to another.

2.1 One-host parasites

2.1.1 Infection by ingestion

Most one-host parasitic protozoa enter the host by being swallowed. This entails protection from desiccation while in the external environment and from destruction by the acidic gastric juice while passing through the stomach. These two problems have been overcome by the development of resistant cysts enclosing a quiescent stage of the parasite. Cysts may be relatively simple proteinaceous structures such as those surrounding parasitic amoebae (e.g. *Entamoeba invadens*, Fig. 3–1) and flagellates (e.g. *Giardia lamblia*), or more complex many-layered structures like those protecting intestinal coccidia such as *Eimeria* species and *Toxoplasma gondii* (Fig. 2–1). Coccidians have a double

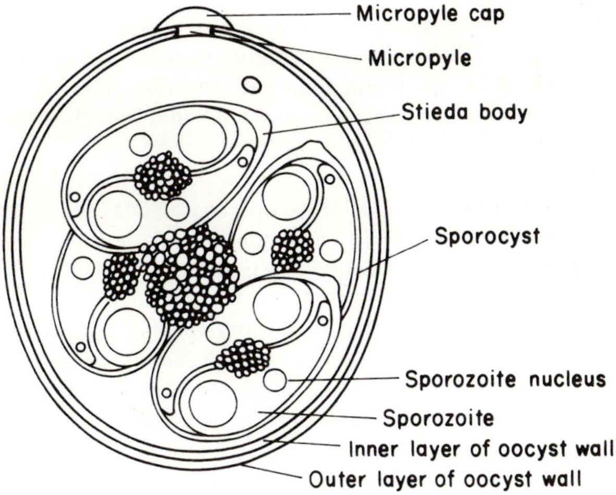

Fig. 2–1 Diagram of structure of oocyst of *Eimeria* (after Levine, 1973, *Protozoan Parasites of Domestic Animals and of Man*, 2nd ed. Burgess Publishing Co., Minneapolis.)

cyst – an outer oocyst containing two or more sporocysts; within the latter lie the infective sporozoites. The oocyst wall has two layers: the outer is a quinone-tanned protein and the inner is protein plus lipid. The resistance of this wall is shown by the fact that it is impermeable to aqueous solutions such as 2.5% potassium dichromate, a disinfectant in which the oocysts are often stored for many months for subsequent experimental use; the dichromate kills the bacteria present in the faeces containing the oocysts but, as it does not penetrate the oocyst wall, has no effect on the contained sporocysts or oocysts. Lipid solvents and gases such as ammonia, oxygen and carbon dioxide can, however, penetrate the wall. The oocyst has an aperture (micropyle), normally secured by a plug. After being swallowed, the cyst is either mechanically broken, for example, in the gizzard of a bird, or the plug is dissolved, perhaps by an enzyme secreted from within after stimulation by prolonged exposure to a high concentration of carbon dioxide in an animal's gut, at about 37°C. The liberated sporocysts are then unplugged (the plug is called the Stieda body) by a series of events involving bile salts and trypsin, in an alkaline environment at about 37°C, and the sporozoites actively wriggle out.

The cyst wall of *Entamoeba* is simpler, about 0.5 μm thick and made of a carbohydrate-protein complex. Excystation (hatching) seems to depend on active movement by the enclosed amoeba, stimulated by an alkaline environment deficient in oxygen such as occurs in a mammal's duodenum. Excystation of the flagellate *Giardia* occurs when the cyst, having been exposed briefly to very acid conditions (pH about 0.5 to 2, for 5 to 30 minutes), subsequently enters a neutral environment.

Thus none of these parasites would emerge from the cyst after being swallowed until the relatively friendly environment of the small intestine had been reached. The amoeba *Acanthamoeba castellanii*, which sometimes lives within human throats, secretes an alkaline protease to aid in its emergence from the cyst, but what stimulates it to do this is unknown. Presumably, as the amoeba is often free-living, the stimulus cannot involve the physiological conditions found in the mammalian gut; and this, in turn, may be why the parasite infects the nasopharynx, trachea and bronchus rather than the alimentary canal.

Not all parasites transmitted by ingestion need a cyst: *Pentatrichomonas hominis*, a harmless commensal of the human gut, does not encyst but succeeds in infecting up to 32% of some human populations. The usual textbook statement that it survives by being 'resistant' merely disguises our ignorance of how it does this.

One-host parasites which infect their hosts by being ingested rely on the host's unhygienic habits to convey them to its mouth, since their cysts leave the host in the latter's faeces. Food, fingers or water are common routes to a new host; many mammals also clean their offspring, or one another, by licking the anal region. Flies or cockroaches may assist in the process. Many one-host parasites have invertebrate hosts (insects and others), which they infect by being ingested, but even less is known of their physiology than our rather meagre knowledge of those with vertebrate hosts.

2.1.2 Infection by other routes

This is rare among one-host parasites. Noteworthy examples include *Trichomonas vaginalis*, which inhabits the human vagina and urethra, and *Tritrichomonas foetus*, which lives in the same organs of domestic cattle. (The spelling of the latter's generic name is not a misprint: most students of the group place the two organisms in separate genera, though both used to be referred to as *Trichomonas*.) Both these species are transmitted during coitus and (like the related *P. hominis*, referred to in section 2.1.1.), neither forms a cyst. *Trypanosoma equiperdum*, a parasite of horses in South America, Soviet Russia, Iran and probably North and South Africa, also relies on its hosts' sexual behaviour for transmission. The parasites live mainly in fluid-filled patches beneath the skin of the penis and vulva and are freed by abrasion of the skin during sexual intercourse. The gregarine *Monocystis* is also, probably, transmitted during sexual union of its earthworm hosts. Location of a new host is made easy for such venereally transmitted parasites. Two protozoa living harmlessly in the human mouth, *Trichomonas tenax* and *Entamoeba gingivalis*, also achieve transmission without needing a cyst, presumably during direct mouth-to-mouth contact (i.e., kissing).

2.2 Two-host parasites

Host location is made easier for these protozoa than it is for most of the one-host parasites, since they usually rely on the activity of one host (the annelid or arthropod vector) to find the other (vertebrate). However, the complexity of adaptation required to adjust their life cycles to those of two very different hosts must make them vulnerable to behavioural changes by either host – or to artificial human interference such as control of vector insects. It is this complexity which is largely responsible, also, for making these parasites such fascinating objects of study.

2.2.1 Infection by ingestion (carnivorism)

Probably quite a common route of entry to the vertebrate host is by the latter eating an infected vector. The parasites are released from the crushed prey in the predator's mouth and penetrate the buccal mucosa – sometimes perhaps through small scratches caused by the hard, chitinous exoskeleton of an ectoparasitic arthropod.

Many species of *Trypanosoma* can be transmitted in this way. Examples of those for which it is the normal route of infection include *T. lewisi* (of *Rattus* spp. and their fleas *Xenopsylla cheopis* and *Ceratophyllus fasciatus*) and probably other trypanosomes of the same subgenus (*Herpetosoma*), *T. corvi* of Corvidae (rooks, etc.) and louse flies (*Ornithomyia avicularia*), and *T. grayi* of African crocodiles (*Crocodilus niloticus*), transmitted by *Glossina palpalis*, a tsetse fly which lives dangerously by sucking blood from within the wide-open mouth of crocodiles basking in the sun and 'panting', like dogs, to cool themselves. The human pathogen *T. cruzi* in South and Central America may well be transmitted thus to some of its rodent and other reservoir hosts, if they eat infected vectors (hemipteran bugs of the family Reduviidae). All these examples are trypanosomes which complete their development in the vector's hind gut (stercorarian development); probably, except for some trypanosomes of fish which are transmitted by ingestion of leeches (Hirudinea), this kind of development occurs in all groups of trypanosomes normally transmitted by ingestion of the vector. Other parasites, not normally transmitted carnivorally, may sometimes be. *Trypanosoma brucei* and *T. congolense*, of African ungulates, are normally transmitted by the bite of infected tsetse flies (*Glossina*: see section 2.2.4), but lions (*Panthera leo*) and other carnivores may become infected by eating infected vertebrate prey. *Toxoplasma gondii* may gain access to its hosts if they ingest infected prey (see sections 3.2.2 and 6.2).

2.2.2 *Infection by ingestion (coprophagy)*

This is really an adjunct of the previous type. *Rattus* species and other rodents may ingest faeces of fleas (Siphonaptera) containing infective metacyclic trypomastigotes of the trypanosome subgenus *Herpetosoma* while cleaning their fur, instead of eating the insects themselves. Birds can become infected similarly with *T. corvi* while cleaning their feathers. *Crocodilus niloticus* can presumably become infected if a defaecating *Glossina palpalis* deposits trypomastigotes of *T. grayi* while, or just after, feeding in the reptile's mouth. Other examples may await discovery.

2.2.3 *Infection by ingestion (haematophagy)*

Almost all two-host parasites infect their invertebrate vectors in this way, by being ingested with the vector's meal of blood taken from an infected vertebrate. The few exceptions include parasites such as *Leishmania tropica* which, as far as is known, does not circulate in the blood of its vertebrate host (man and other mammals), but is confined to a single sore on the skin. *L. tropica* presumably therefore relies, in order to get into its vector (*Phlebotomus* sp.), on a hungry insect probing through an infected area of skin when seeking its blood meal. Some parasitic protozoa have special stages in the vertebrate's blood destined to infect the vector. These may be sexual stages (gametocytes) in malaria parasites (*Plasmodium* spp.) and adeleine haemogregarines. Morphologically and physiologically differentiated asexual individuals play this role in some, if not all, species of *Trypanosoma* and, probably, some eimeriine haemogregarines (*Lankesterella, Schellackia*), the sporozoites of which are destined for passive transfer to a new vertebrate host by an ectoparasitic mite (Acarina). Other parasites are not known to produce special forms to infect vectors (e.g. *Leishmania* and *Babesia*). However, this assertion may merely reflect current ignorance (especially with *Babesia*).

Protozoa have not generally developed the degree of synchronization between their life cycles and their vector's feeding behaviour that certain filarial nematodes have, though something similar seems to have been developed by certain species of *Plasmodium*, the gametocytes of which mature at the times of their vectors' greatest feeding activity. This usually occurs in the evening or at night, the vectors being female anopheline mosquitoes. Some species of *Trypanosoma* may be commoner in their hosts' peripheral blood at times when vectors are more likely to feed (see section 6.2).

2.2.4 *Infection by penetration of skin and other membranes*

Unaided penetration of intact skin is probably beyond the power of minute parasitic protozoa. However, penetration through possibly damaged buccal mucosa probably occurs (sections 2.2.1 and 2.2.2). *Trypanosoma cruzi* (which causes Chagas's disease of human beings in South and Central America) may penetrate the conjunctiva of a vertebrate host (including man) if deposited on or near it by a defaecating infected vector (Hemiptera, Reduviidae). Surprisingly, few protozoa can cross undamaged placentas. *Toxoplasma gondii*, one which can (too often with tragic results), presumably does so as a passenger within wandering macrophages (see section 3.2.2). However, *Leishmania*, which also inhabits macrophages, rarely if ever crosses the placenta. Many two-host protozoa use their feeding vector's proboscis to penetrate the skin of the vertebrate host. Species of *Plasmodium, Leishmania, Trypanosoma* (those with salivarian development), *Babesia, Theileria, Haemogregarina* have developed life cycle specializations to facilitate such assisted penetration; all complete their development in the vector's salivary glands (*Plasmodium, Trypanosoma brucei,*

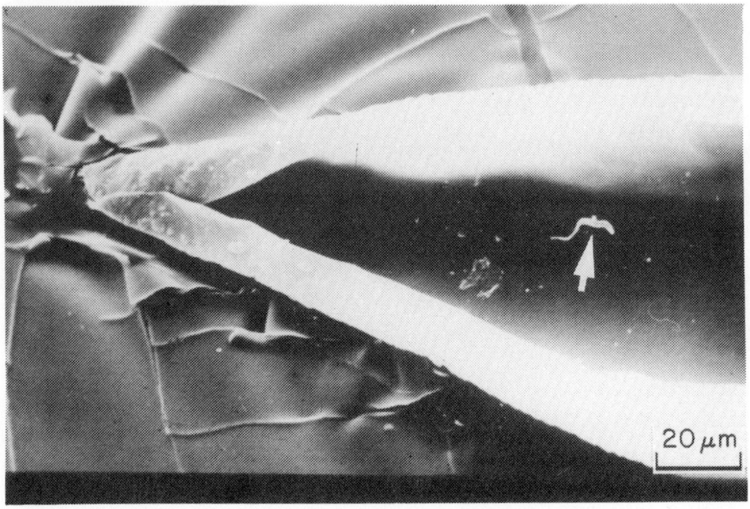

Fig. 2–2 Scanning electron micrograph showing *Trypanosoma brucei* (arrowed) in tip of proboscis (labrum) of *Glossina morsitans*. (From Molyneux, 1980, *Insect Sci. Application*, **1**, 39.)

Babesia, Theileria), proboscis (*Trypanosoma congolense, T. vivax*) or anterior midgut (*Leishmania* of mammals), as described in section 3.2 (see Fig. 2–2).

Another remarkable adaptation has been developed by *Trypanosoma brucei* to ease transmission from its insect host *Glossina* to its mammalian hosts. Although the trypanosome achieves infection of the salivary glands in only a small proportion of infected *Glossina*, the presence of parasites attached to the insect's proboscis wall interferes with receptors which measure the blood flow while the *Glossina* is feeding (Fig. 2–3). This causes the insect to feed more voraciously and more frequently than normal, thereby increasing the chances of infective trypanosomes from the salivary glands being inoculated into the unfortunate mammal, and may also lead to increased longevity of the insect.

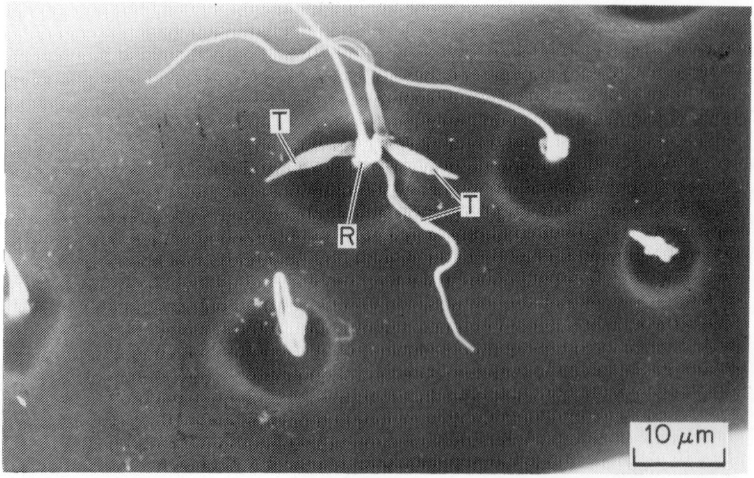

Fig. 2–3 *Trypanosoma brucei* (T) attached to mechanoreceptor (R) in labrum of *Glossina morsitans*. (Scanning electron micrograph from Molyneux, 1980, *Insect Sci. Application*, **1**, 39.)

3 Development in the Host (Life Cycles and Physiology)

3.1 One-host parasites

Some parasites of this group have very simple life cycles, but this is not always so. The parasitic amoeba, *Entamoeba histolytica*, has a cycle involving only excystment (section 2.1), asexual binary fission either in the lumen or wall of the large intestine of man and some other mammals (and occasionally in other tissues: section 5.2), and re-encystment (section 6.1); within the cyst the parasite's nucleus divides twice by mitosis. Other gut-dwelling amoebae and flagellates (mostly commensals, though one – *Giardia lamblia* – can be harmful: see section 5.2) have similarly basic life cycles. *Entamoeba histolytica* feeds by a process (phagotrophy) which consists of sucking up the bacteria, food debris or tissue cells which form its prey, into its cytoplasm from its hind end (uroid), which may be specialized for this purpose, although no particular structure can be seen there. The trichomonads and other intestinal flagellates have a cytostome for similar phagotrophic feeding, though none has been described in *Giardia lamblia*. Ingestion of small droplets of liquid (pinocytosis) and transport of simple nutrients across the plasmalemma probably also occur. These dwellers in the intestinal lumen can metabolize without oxygen (i.e., they are facultative anaerobes), though they use oxygen when available. The cysts of *Entamoeba* contain crystalline ribosomal arrays (Fig. 3–1), presumably stacked ready for use by the emergent parasite whose first activity is to divide into small but rapidly growing 'amoebulae'; this probably requires a considerable amount of synthetic activity.

Some parasites inhabiting the gut wall and, sometimes, other organs of a single host species have more complex cycles, involving alternation of asexual and sexual generations (Fig. 3–2). This type of cycle is shown by the Gregarinia and some eimeriine Coccidia, and resembles that of the two-host malaria parasite *Plasmodium* (section 3.2). In the Gregarinia, all of which are parasites of invertebrates (usually insects), sexual union and subsequent division (sporogony) occur within the host. Two sexual individuals (male and female gametocytes) become associated (syzygy) and secrete a cyst wall (gametocyst) around themselves. They then mature into gametes (gametogony), without any reduction division (meiosis), the parasite being haploid for most of its life cycle, and fertilization occurs. Subsequent nuclear and cytoplasmic divisions (sporogony) – the first nuclear division being meiotic – produce many small sporozoites, the forms which infect the next host. A few Gregarinia also undergo a process of multiple asexual division called schizogony, but this has been dropped from the life cycle of most.

Many one-host eimeriine Coccidia are important pathogens of domestic

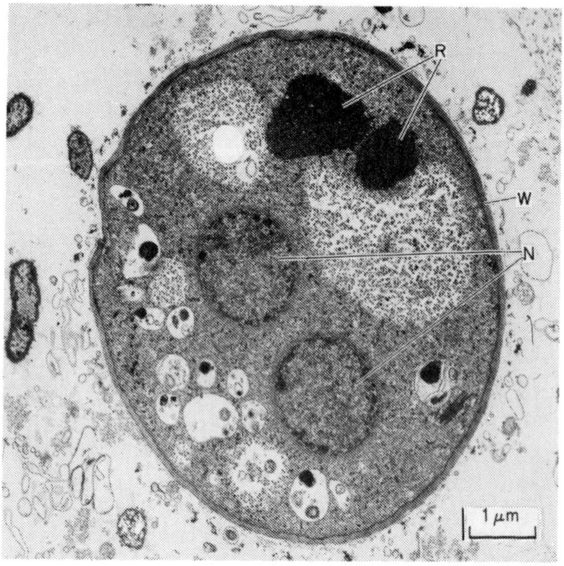

Fig. 3–1 Thin section of cyst of *Entamoeba invadens* showing 2 of the 4 nuclei (N), ribosomal helices in crystalline arrays ('chromatoid bodies' – R) and cyst wall (W). (Micrograph by D. C. Barker, Molteno Institute, Cambridge.)

animals, particularly birds; most of these belong to the genus *Eimeria*. After a bird swallows an infective, resistant oocyst, the contained sporozoites escape into the lumen of its small intestine (section 2.1.1) and penetrate the plasmalemma of their host cells (often the cells of the adjacent gut wall) by active movement probably aided by secretions from the organelles (rhoptries and micronemes; Fig. 3–3) in the apical complex at their anterior end. (It is because of the possession of this typical apical complex that the phylum containing these, and other, parasites is now named the Apicomplexa.) Some coccidia enter macrophages first and allow these wandering cells to transport the sporozoites to their final host cells, which are in the walls of the intestinal caeca of chickens for *Eimeria tenella*, or the bile duct epithelium of rabbits for *E. stiedai*. Within the final host cells, the parasites grow larger (trophozoites) and then undergo schizogony, a process of asexual multiplication typical of, if not restricted to, the Apicomplexa. (There is controversy about whether schizogony also occurs in the enigmatic phylum Microspora.) Essentially, in schizogony, two to many repeated mitotic nuclear divisions are followed by cytoplasmic cleavage into as many small uninucleate organisms (merozoites) as there were nuclei in the unsegmented schizont which, during this cleavage, may be called a segmenter. The merozoites, which have an apical complex similar to that of the sporozoites, escape from the damaged host cell. They enter other nearby cells to repeat the process (schizogony) for one or a limited number of further generations. Eventually, for reasons unknown, merozoites arise which do not undergo

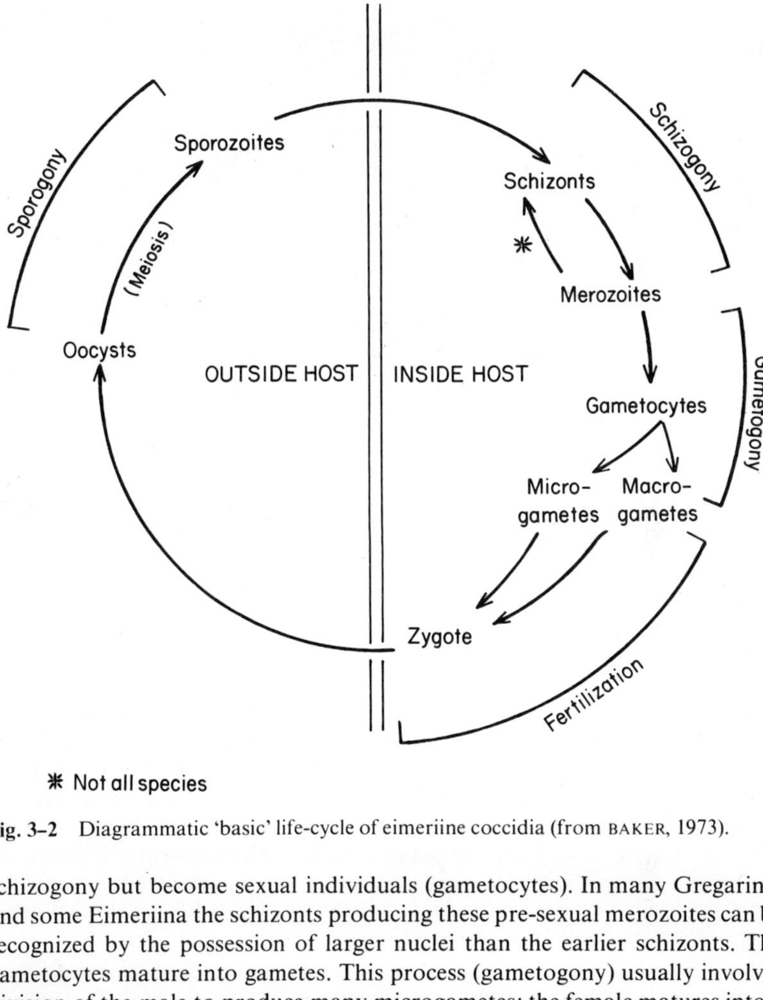

Fig. 3–2 Diagrammatic 'basic' life-cycle of eimeriine coccidia (from BAKER, 1973).

schizogony but become sexual individuals (gametocytes). In many Gregarinia and some Eimeriina the schizonts producing these pre-sexual merozoites can be recognized by the possession of larger nuclei than the earlier schizonts. The gametocytes mature into gametes. This process (gametogony) usually involves division of the male to produce many microgametes; the female matures into a single macrogamete except in Gregarinia, where both sexes divide within the gametocyst (see above). After fertilization, which with most eimeriines occurs in the lumen of the host's intestine, the resultant diploid zygote encysts within an oocyst. The zygote undergoes reduction division (meiosis) followed by one or more mitotic divisions to form four or more sporozoites; most Gregarinia and all species of *Eimeria* form eight. For greater security, the sporozoites of *Eimeria* and many other eimeriines are enclosed (in pairs in *Eimeria*) within a second resistant cyst, the sporocyst (the eponymous 'spore' of the phylum's older name Sporozoa). Sporozoite formation (sporogony) occurs either just before, during or just after exit of the oocyst from the host (in faeces, in the case of all eimeriines). The sporozoites then rest within the spore until the latter is

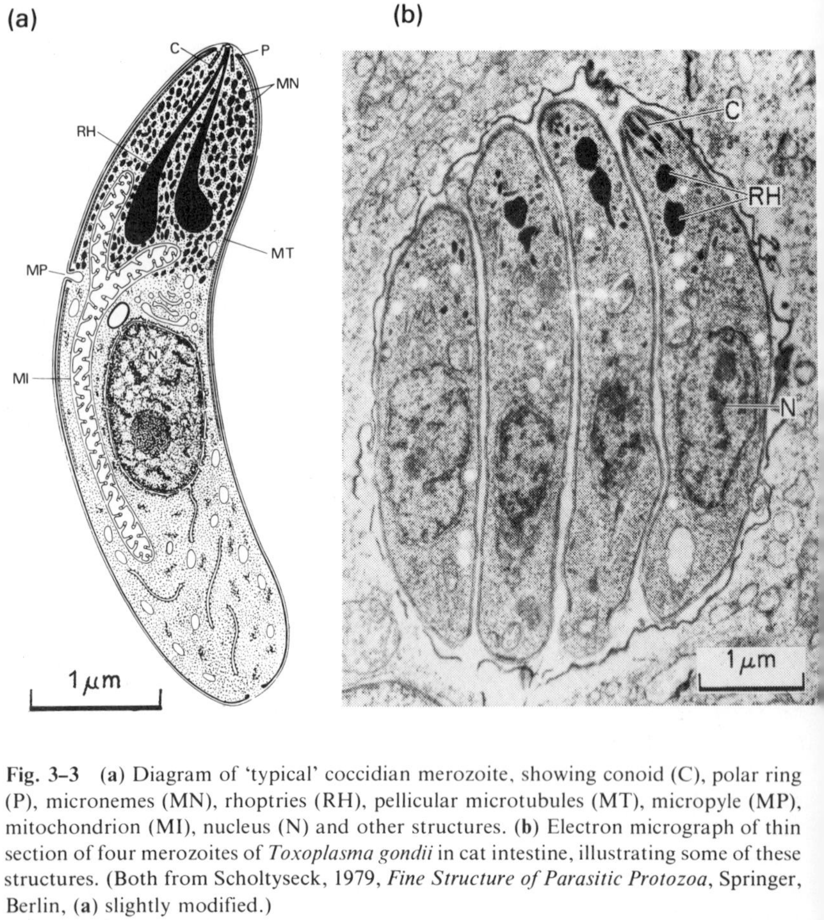

Fig. 3–3 (**a**) Diagram of 'typical' coccidian merozoite, showing conoid (C), polar ring (P), micronemes (MN), rhoptries (RH), pellicular microtubules (MT), micropyle (MP), mitochondrion (MI), nucleus (N) and other structures. (**b**) Electron micrograph of thin section of four merozoites of *Toxoplasma gondii* in cat intestine, illustrating some of these structures. (Both from Scholtyseck, 1979, *Fine Structure of Parasitic Protozoa*, Springer, Berlin, (**a**) slightly modified.)

swallowed by a new, suitable host (section 2.1.1). This cycle, truly a 'tricycle' of schizogony, gametogony and sporogony, is at first sight alarmingly complex; it is interestingly modified by the Sporozoea which have adopted a two-host life style (section 3.2). The occurrence of meiosis during the first nuclear division of the zygote results in these organisms, and probably all Apicomplexa, being haploid during most of their life cycle: only the zygote is diploid.

Little is known of the physiology of the parasites described in this section. Probably the intracellular trophozoites feed by phagotrophy, as does *Plasmodium* (see section 3.2). VON BRAND (1973) wrote that 'Sporozoa are essentially aerobic'; surprisingly, in view of their habitat in the gut lumen or faeces, eimeriine oocysts need oxygen for sporogony, though they can survive without it. Oocysts of some eimeriines contain a probable food reserve ('residual body'), which is consumed by the enclosed organism while undergoing sporogony.

3.2 Two-host parasites

There are two major groups of protozoa with representatives in this category – Apicomplexa and Mastigophora. The basic life cycle of the former has been described in section 3.1 (Fig. 3–2); it remains only to discuss its modification by the species which have adopted an additional host. The second host is usually an arthropod vector, as with malaria parasites (Haemosporina) and Piroplasmia, but may be another vertebrate (as with *Toxoplasma*: see section 3.2.2).

3.2.1 Haemosporina and Piroplasmia

From the diagrammatic life cycle of the Haemosporina in Fig. 3–4, it is clear that this group has simply transferred the later stages of gametogony, fertilization and sporogony to the secondarily acquired vector host; in the basic sporozoan cycle (Fig. 3–2), most of this sequence occurs either in the lumen of the vertebrate's gut or in the outside world. This transfer having conferred protection upon the sporogonic stages of Haemosporina, and the need for a resistant oocyst having therefore been eliminated, this stage has become a thin-walled expandible cyst within the body of the vector insect (on the outer surface of its midgut). Inside this oocyst, many sporozoites develop, no longer contained within sporocysts. After escaping from the ruptured mature oocyst, the sporozoites find their way to the vector's salivary glands. They pass through the single layer of secretory cells forming the wall of the paired glands, and thus are delivered directly into the blood stream of the next vertebrate host on which the infected insect feeds. (Saliva is injected at the time of biting, as an anticoagulant.) Members of the Haemosporina, therefore, no longer develop a true 'spore' (=sporocyst). One genus of the suborder, *Plasmodium* (the malaria parasite), has specialized further by developing, within its vertebrate host, an additional cycle of schizogony. This cycle, comprizing a potentially unlimited number of generations, occurs within the vertebrate's red blood cells (erythrocytes); typically, its occurrence is remarkably synchronized, with a period of 12–72 hours depending on the species (see sections 5.2.2 and 6.2). In all species of *Plasmodium* except those infecting mammals, the phylogenetically 'earlier' schizogonic cycle in fixed tissue cells is also potentially capable of unlimited repetitions, unless of course the parasites are eliminated by the host's immune response or by drug treatment. In the species infecting mammals (including ourselves), however, this exoerythrocytic schizogony, which occurs only within liver parenchyma cells (hepatocytes), seems to be restricted to a single cycle. There is still doubt about precisely what does occur, but it is now thought that, in two of the species of *Plasmodium* which infect man (*P. vivax* and *P. ovale*; see appendix), two phenotypes or genotypes of sporozoite may exist. One undergoes a single cycle of schizogony in hepatocytes immediately after its inoculation by the biting vector. The other type (a hypnozoite) perhaps remains quiescent ('sleeping' – hence the name 'hypnozoite') until reactivated by some unknown endogenous or exogenous stimulus months or even years later. Reactivation would result in a typical relapse of malaria. In related parasites of birds

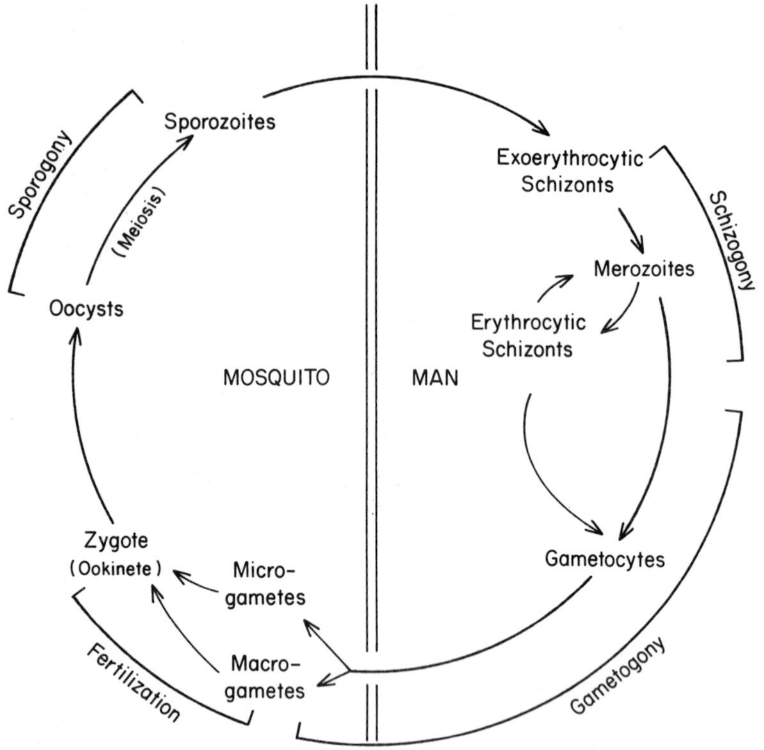

Fig. 3–4 Diagrammatic representation of the life cycle of *Plasmodium* spp. (modified from BAKER, 1973).

(*Leucocytozoon* spp.), which do not undergo schizogony within their hosts' erythrocytes, the stimulus reactivating the hypothetical hypnozoites seems to be related to the onset of the birds' breeding behaviour, either directly or via a common initiating event such as increasing day length.

The Piroplasmia (*Babesia* and *Theileria*) have simpler life cycles, though not all details are known. *Babesia* infects only the red blood cells of its vertebrate hosts (mainly Bovidae and Rodentia, though rare – but often fatal – accidental human infections occur, often in people whose spleens have been surgically removed). In the red blood cells, it divides by schizogony into four merozoites. The vectors of *Babesia* spp. are ticks (Acarina). In the ticks a series of asexual multiplications occurs, culminating in infection of the salivary glands of subsequent stages in the tick's life cycle. Transmission can occur from the mother tick to its larvae, via the ovary (trans-ovarial), or from larva to nymph or nymph to adult (trans-stadial). *Theileria* multiplies in its vertebrate hosts (mainly Bovidae) by schizogony in fixed tissue macrophages. (Before their true nature was known, these schizonts were called 'Koch's blue bodies'.) Merozoites enter red blood cells but do not divide in them. The vectors are again ticks, in

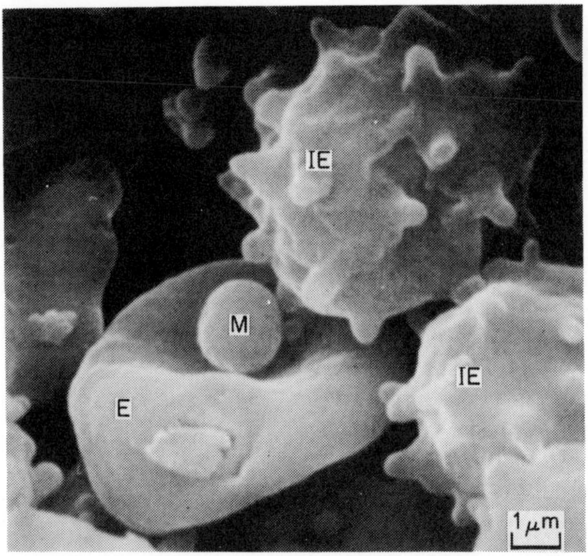

Fig. 3–5 Scanning electron micrograph showing merozoite (M) of *Plasmodium knowlesi* attached to an erythrocyte (E) of *Macaca rhesus* in vitro, and two other erythrocytes (IE) containing schizonts (from Bannister, 1977, *Symp. Brit. Soc. Parasitol.*, **15**, 27).

which an apparently simple but not fully known development occurs. Transstadial, but not trans-ovarial, transmission can occur. Finally the tick's salivary glands are infected by tiny parasites. Sexual processes have recently been reported, but need confirmation.

Quite a lot has been discovered about the way in which *Plasmodium* merozoites enter erythrocytes. The merozoite becomes attached by its apical complex (Fig. 3–3) to specific receptors on the red cell's membrane (Fig. 3–5). The receptors on human erythrocytes for *P. vivax* and *P. knowlesi*, though not for *P. falciparum*, are closely associated with the Duffy blood-group antigen molecules. (*P. vivax* and *P. falciparum* are both predominantly parasites of man, whereas *P. knowlesi* naturally infects *Macaca rhesus*; it can, however, infect man also.) Attachment to the receptor triggers release of some substance from the merozoite's rhoptries (Fig. 3–3), which causes the red cell membrane to invaginate, drawing the attached merozoite into the cell. The membrane then reseals across the hole left at the site of intucking. The parasite is thus now contained within a vacuole lined with erythrocyte membrane. The related genus *Babesia* (Piroplasmia) probably enters red blood cells similarly, but somehow the enveloping membrane disappears and the parasite soon lies free within the erythrocyte cytoplasm.

Once within the red blood cell, *Plasmodium* begins to feed through a small cytostome. The parasite is bounded by a two-layered pellicle, except at the base of the cytostome, where there is only a single unit membrane. Here the parasite engulfs 'mouthfuls' of host cytoplasm, including the membrane lining the surrounding parasitophorous vacuole. This process is called phagotrophy. The food vacuole (phagosome) within the parasite is pinched off, passes into the parasite's cytoplasm and fuses with a lysosome in the conventional way. Within the resultant secondary lysosome, acid phosphatase and other hydrolytic enzymes digest the erythrocyte cytoplasm. The indigestible remains of the haemoglobin, including the iron, some of the globin and possibly some parasite proteins, are seen as brown or black 'malarial pigment' (haemozoin) within the vacuole, now a postlysosome. These packets of pigment (the parasite's 'faeces') are discarded when the schizont ruptures (a process reminiscent of the waste disposal technique used by lunar astronauts!). *Babesia* and *Theileria* presumably ingest and digest their food similarly, but apparently break down the haemoglobin further since no visible undigested pigment remains.

Surprisingly, since they live in an environment rich in oxygen, the intra-erythrocytic stages of malaria parasites of primates and murine rodents lack structurally typical mitochondria and apparently respire largely anaerobically. Even the species infecting birds, which have recognizable mitochrondria and a functional Kreb's cycle, Embden-Meyerhof pathway and cytochrome chain enzymes, rely considerably on anaerobic glycolysis as an energy source. The stages in the vector of all species of *Plasmodium*, as far as is known, use conventional aerobic respiratory pathways. (This behaviour can be compared with that of the trypanosomes; see section 3.2.3.) Intraerythrocytic *Babesia* and *Theileria*, and stages of *Babesia* in the vector, also lack typical mitochondria. The blood stages of *Babesia*, at least, metabolize glucose to lactate very largely by anaerobic pathways, though they do use some oxygen.

3.2.2 Toxoplasma *and its relatives*

Toxoplasma gondii has already been mentioned as an eimeriine coccidian which has adopted a secondary host species – in this instance, another vertebrate. *T. gondii* has not done this by inserting the originally extra-hostal part of its life cycle into another host (as the Haemosporina have, into an insect; section 3.2.1). Instead, it has developed an extensive extra phase of asexual division and encystment within the tissues of the secondary host. The second host may be any species of warm-blooded vertebrate (bird or mammal, including human beings). The primary (or definitive) host is the domestic cat (*Felis domesticus*) or one of a few closely-related felids. The original schizogonic-gametogonic cycle occurs in the intestinal wall of the cat. This typically eimeriine phase was completely unknown until 1969, when it was discovered almost simultaneously by Frenkel in the U.S.A. and Hutchison in Scotland. Until it was known, the correct taxonomic position of *Toxoplasma* was undecided: it had even been classified as a fungus. Sporogony in the cat results in oocysts characteristic of those produced by the eimeriine coccidian *Isospora*. Even twelve years later, the nomenclatural confusion resulting from this discovery has not been sorted out. The situation is further confused because many erstwhile 'species' of *Isospora* are now known to be stages in the life cycles of parasites related to *Toxoplasma*, and previously only imperfectly described. They are currently being distributed between the 'genera' *Sarcocystis, Besnoitia, Hammondia, Levinea* and *Cystoisospora*, but the last word has most certainly not yet been said.

The 'additional' exo-enteric multiplicative phase of *T. gondii* occurs in macrophages. Parasites structurally resembling merozoites, known as endozoites or tachyzoites, divide by endogeny ('internal budding') within a parasitophorous vacuole or 'pseudocyst'. After their release, endozoites may re-enter macrophages, probably aided by secretions from their anterior organelles (compare *Plasmodium*, section 3.2.1), to recommence the cycle. Alternatively, they may enter neurones or muscle cells and encyst. Within these tissue cysts (so-called to distinguish them from oocysts), they again multiply by endogeny to form several hundred cystozoites (bradyzoites). The tissue cysts are remarkably resistant, and can be transferred to new hosts (including cats and human beings) by the consumption of raw or undercooked infected meat. One notorious small epidemic was traced to an overworked hamburger stall at an American university (see section 6.2). Endozoites can cross the placenta and infect the foetus of a woman suffering from acute toxoplasmosis. This does not occur if the pregnant woman is chronically infected, when her immune response has limited the parasite to the non-pathogenic tissue cysts. Human beings, cats and other hosts can also become infected by the conventional eimeriine means of oocyst ingestion; this is probably the commonest route.

3.2.3 Trypanosomatidae

The life cycles of *Leishmania* species are relatively simple. All inhabit only macrophages in their vertebrate hosts (a few lizards, many mammals), where they multiply by binary fission as amastigotes (sometimes now called micromastigotes as they possess a rudimentary flagellum; see Fig. 3–6). In their vectors, sandflies (Diptera, Psychodidae, Phlebotominae), they elongate to become promastigotes and again reproduce by binary fission, initially in the midgut. The few species which infect lizards then move to the hindgut wherein they complete their so-called hypopylarian life cycle. All subspecies of *L. brasiliensis* (which infect mammals in South America; see appendix) move first to the hindgut and then to the insects' foregut to complete their (peripylarian) development. Species of the *L. mexicana, L. donovani* and *L. tropica* complexes move directly from midgut to foregut (suprapylarian development; 'pylarian' in these compound words refers to the pyloric region separating the sandfly's midgut from its hindgut). In the vertebrate hosts, cells infected with parasites of the *L. brasiliensis, L. mexicana* and *L. tropica* groups are restricted to the skin, possibly because the parasites cannot survive at the slightly higher temperature of the internal organs. Cells infected with *L. donovani* or its close relatives are found throughout the mammalian body, predominantly in liver, spleen and bone marrow.

Life cycles of almost all species of *Trypanosoma* involve two hosts, a vertebrate and an invertebrate. In the former, the organisms exist as extracellular trypomastigotes, inhabiting blood plasma or other tissue fluids;

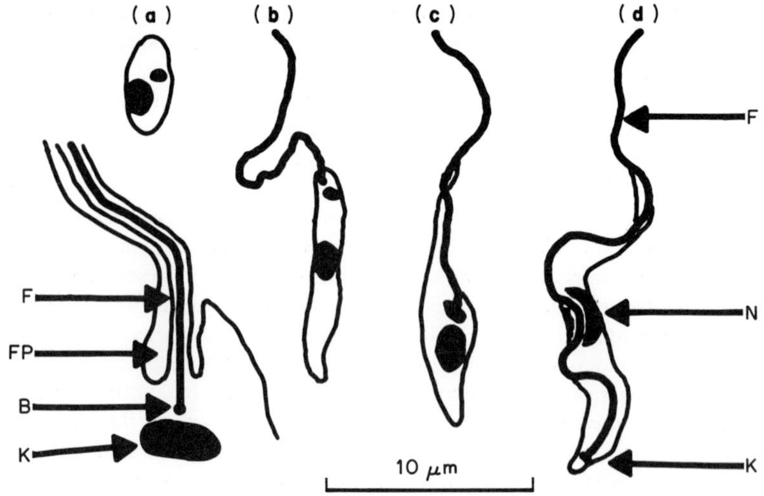

Fig. 3–6 Diagram showing cellular organization of the various morphae which may occur in the life cycles of species of *Leishmania* or *Trypanosoma*. (**a**) amastigote (micromastigote), (**b**) promastigote, (**c**) epimastigote, (**d**) trypomastigote. B, basal body; F, flagellum; FP, flagellar pocket; K, kinetoplast; N, nucleus. (Slightly modified from Baker, J.R., 1978, in *Companion to Microbiology*, Bull and Meadow, Longman, London.)

members of the subgenus *Schizotrypanum* replicate as amastigotes within fixed tissue macrophages or muscle cells. In their invertebrate vectors, *Trypanosoma* species commence development in the midgut and complete it in either the hindgut (stercorarian development), as most species do, or in the foregut (including sometimes the salivary glands), as do only species of the subgenera *Trypanozoon, Nannomonas* and *Duttonella* (see appendix) and some parasitizing fish and leeches (Annelida, Hirudinea); the latter type of development is called salivarian (this adjective, and stercorarian, are derived from the Greek words for 'saliva' and 'faeces' respectively, referring to the medium in which the trypanosomes end their development). Both types of development in the vector consist essentially of binary fission by extracellular epimastigotes; in the species with salivarian development, this is prefaced by a phase of binary fission as 'procyclic' trypomastigotes in the vector's (*Glossina* species) midgut. The epimastigotes finally differentiate into non-reproducing 'metacyclic' trypomastigotes (infective to the vertebrate host) in the vector's hindgut (stercorarian development) or in its proboscis or salivary glands (salivarian development). The accepted view is that those species (subgenus *Trypanozoon*) which invade the salivary glands reach them by migrating from the midgut forwards down the proboscis and then returning up the hypopharynx to the glands. Recently the occurrence of this rather improbable odyssey has been questioned and some evidence produced that the parasites penetrate the midgut wall (apparently passing through its component cells) and reach the salivary glands via the insect's haemocoel.

In the subgenus *Trypanozoon*, and probably in *Nannomonas* and *Duttonella*, the surface of the trypomastigotes living in the blood plasma is covered by a layer of glycoprotein, secreted by the parasites, which forms a surface antigen whose amino acid sequences are variable, and this enables the parasites to survive attack by the host's antibodies (see section 4.1); there is some evidence that host plasma proteins may be adsorbed to the trypanosomes' surface in the subgenus *Duttonella*. This surface coat (variable surface antigen) is lost soon after ingestion by the vector (*Glossina*); in the subgenera *Trypanozoon* and *Nannomonas* it is redeveloped by the metacyclic trypomastigotes in the salivary glands or proboscis, respectively. In *Duttonella* it seems that the surface coat is not re-acquired until after re-entry into a vertebrate host; this, plus the observation that hostal proteins may be adsorbed onto the surface of species of *Duttonella*, suggests that the latter's coat may be rather different from that of the other two subgenera. In members of all three subgenera with salivarian development in *Glossina* (i.e., *Trypanozoon, Nannomonas* and *Duttonella*) there are profound differences in the respiratory mechanism of the different forms (morphae) developing during the life cycle. These differences have been studied most fully in *Trypanozoon*. The trypomastigotes which inhabit the vertebrate host's blood (haematozoic trypomastigotes), and the metacyclic trypomastigotes, have only a single, structurally reduced non-functional mitochondrion. Their highly aerobic respiration follows at first the Embden-Meyerhof glycolytic pathway as far as pyruvate, excreted as pyruvic acid, but terminal respiration involves an L-alpha-glycerophosphate oxidase/dehydrogenase system located in small membrane-bound 'hydrogenosomes' in the cytoplasm. The procyclic trypomastigotes and epimastigotes within the vector also have only a single mitochondrion, but it is elongated and fully-developed structurally, with a functional Kreb's cycle and cytochrome chain; these forms therefore fully degrade glucose to carbon dioxide and water. It is speculated that the 'switching on and off' of the Kreb's cycle and cytochrome chain enzymes is controlled by the unusually large mass of intramitochondrial deoxyribonucleic acid (the kinetoplast) characteristic of Trypanosomatidae and other members of the order Kinetoplastida. Certainly no other function has yet been convincingly ascribed to this organelle, the DNA of which consists of circular molecules of two sizes – minicircles (about 5500 of them in *T. brucei*, each composed of some 1000 nucleotide base pairs) and maxicircles (about 45, consisting of 20 800 base pairs, in *T. brucei*) (Fig. 3–7). The maxicircles are probably equivalent to conventional eukaryotic mitochondrial DNA, coding for enzymes of the Kreb's cycle and cytochrome chain. The function of the minicircles is unknown; possibly they control the 'switch', but it is puzzling why so many small molecules should be needed – and why they should also occur, as they do, in species lacking a complete 'switch' (see below). Probably the 'stumpy' haematozoic trypomastigotes (Fig. 3–8a), whose main function is generally believed to be development in *Glossina* rather than continued replication in the vertebrate host, have a partially functional mitochondrion as a pre-adaptation for the enzymatic 'switching on' required after their ingestion by the vector. The non-mitochondrial respiratory system is wasteful of both oxygen and glucose in

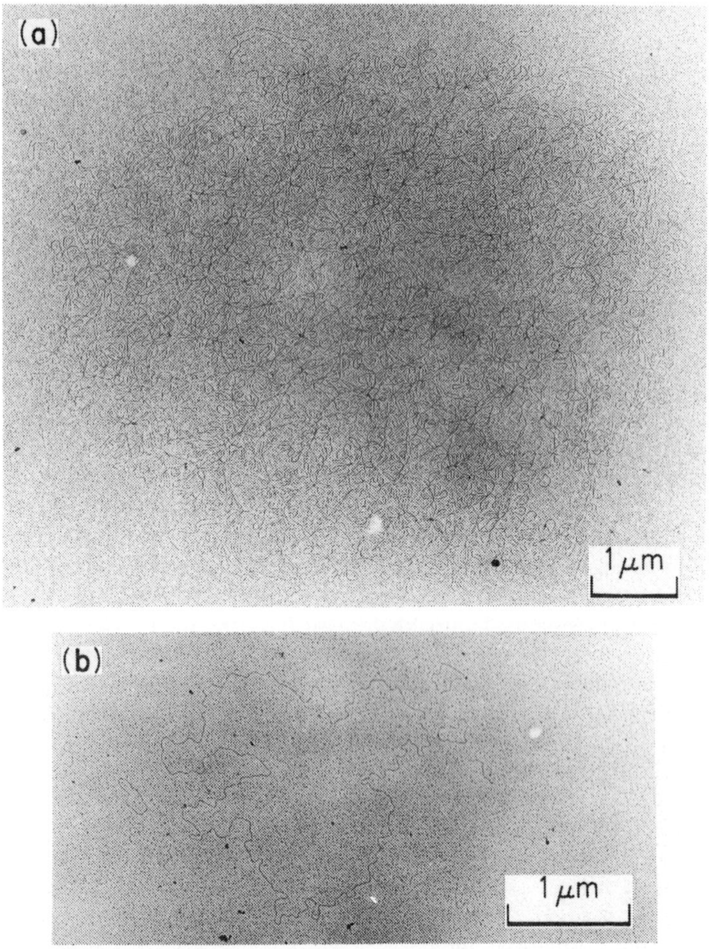

Fig. 3–7 (**a**) Network of DNA molecules from 'exploded' kinetoplast of *Trypanosoma equiperdum*, showing linked 'minicircles' and occasional peripheral 'maxicircles' (from Hajduk and Cosgrove, 1979, *Biochim. Biophys. Acta*, **561**, 1). (**b**) Isolated DNA molecules from kinetoplast of *Crithidia fasciculata* (micrograph by S. J. Hajduk, University of Glasgow).

terms of energy output, but presumably the parasite can afford this in an environment as rich in both components as are blood plasma and tissue fluids. Most other species of Trypanosomatidae which have been investigated (not very many) do not show this 'switch', having more or less fully functional mitochondria throughout their life cycle, and it is not clear what advantage has accrued to those species which have adopted the more wasteful method during part of their life cycle; perhaps it enables them to have a more rapid energy turn-

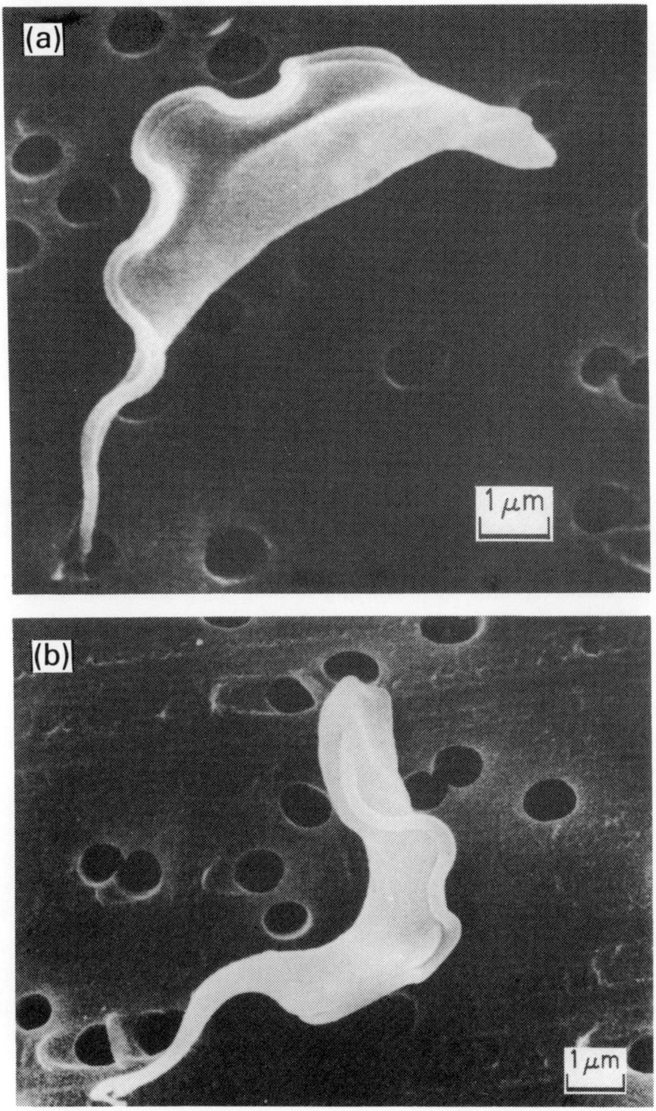

Fig. 3–8 Scanning electron micrograph of (**a**) short, stumpy, haematozoic trypomastigote and (**b**) slender haematozoic trypomastigote of *Trypanosoma brucei*. (Micrographs by L. Tetley, from Vickerman and Tetley, 1977, *Ann. Soc. belge Méd. trop.*, **57**, 441.)

over, related to their rapid rate of replication ('slender' haematozoic trypomastigotes (Fig. 3–8b) have a minimum doubling time of about $3\frac{1}{4}$ hours), shared with few, if any, other species in the family.

4 Outwitting the Host's Defences

There is little, if any, evidence of defensive reactions by invertebrate vector hosts against their protozoan parasites, though one at least (*Trypanosoma rangeli* in *Rhodnius prolixus*) is harmful to the insect (section 5.1). In their vertebrate hosts, however, parasites may be subjected to all manner of immunological and cellular defensive mechanisms (INCHLEY, 1981), and many have evolved sophisticated means of avoiding these, or at least mitigating the effects.

4.1 Extracellular parasites

Protozoa living solely in the gut lumen seem not to stimulate their host antigenically, and so do not produce an immune response. However, when parasites such as *Entamoeba histolytica* invade the intestinal mucosa or other tissue, the host responds by producing serologically detectable antibodies but there is little evidence that these are protective; nor is there any local cellular inflammatory response. There is some experimental evidence that *E. histolytica* can undergo a change of surface antigenic structure under the influence of antiserum, but whether this is of any importance in life is unknown.

Extracellular parasitic protozoa living in the blood (e.g., the trypanosomes) undoubtedly stimulate an immune response by their host. From the parasite's viewpoint, rapid elimination of the infection (like rapid death of the host) is undesirable, as it would reduce the time available for the infection of the vectors on which the parasite depends for transmission to its next vertebrate host; the more vectors are infected, the greater are the chances of the parasite's survival, and dissemination. Trypanosomes avoid rapid elimination in one or more of three ways: by adopting an intracellular habitat for part of their life cycle (subgenus *Schizotrypanum*; section 4.2), by hiding in tissues with minimal blood flow and thus reducing exposure to antibodies, or by changing the structure of their surface antigens at the parasite/host (antigen/antibody) interface.

Trypanosoma lewisi and probably *T. musculi*, cosmopolitan parasites of rats and mice respectively, and both members of the subgenus *Herpetosoma*, induce initially the synthesis by their hosts of an IgG antibody, 'ablastin', which does not contribute to lysis of the parasites but, apparently by interfering with translocation of carbohydrates, thymidine and perhaps other nutrients across the parasites' surface membrane, inhibits their reproduction. This effect is reversible: if parasites are removed from the ablastin-containing ambient by inoculation into a new host, they recommence reproduction. In *T. lewisi* infections, a trypanosomicidal antibody is produced at about the same time as ablastin, and the trypanosome population is markedly reduced – though not to zero. The surviving trypanosomes differ antigenically, presumably in the configuration of their surface proteins (see below), and so are not susceptible to the antibody, though they remain sensitive to ablastin which presumably reacts with some other membrane component. Being present in only small numbers, they do not provide much antigenic stimulation to the host and so survive in its circulation as a source of potential infection to vectors (Siphonaptera) for two or three weeks before the host synthesizes a second trypanosomicidal antibody. Unable apparently to undergo further antigenic change (unlike the salivarian trypanosomes, see below), all the *T. lewisi* are destroyed. Production of trypanosomicidal antibody concurrently with ablastin has not been conclusively demonstrated in *T. musculi* infections, but a lethal antibody appears later and eliminates parasites from the peripheral blood circulation; a few individuals survive, without antigenic change, in the vasa recta of the kidneys; these are blood sinuses presumably constituting a circulatory 'backwater' in which the concentration of plasma antibodies is low, and in which the parasites are less exposed to attack by predatory macrophages. As the host's peripheral circulation remains solidly immune to re-infection, the advantage to the parasites of this survival in the vasa recta, out of reach of hungry vectors, is not obvious. Ablastin production clearly benefits the parasites, as well as the hosts, by limiting parasitaemia to a level which is neither lethal to the host nor capable of providing sufficient antigenic stimulation to lead to rapid synthesis of trypanosomicidal antibody; thus parasites remain available to vectors for longer than they would otherwise do.

The salivarian (*Glossina*-transmitted) trypanosomes have prolonged their survival in peripheral blood and therefore their availability to vectors, by postponing their own and their host's total destruction in way of a more postponing their own and their host's total destruction in a more sophisticated type of surface antigen variation. Haematozoic and metacyclic trypomastigotes (i.e., those inhabiting mammalian blood and *Glossina* salivary glands respectively) of *T. brucei* have a layer of glycoprotein 15nm thick coating the exterior of their cell membrane (plasmalemma) (Fig. 4–1). This layer contains about 7×10^6 molecules (molecular weight about 65 000), containing some 600 amino acid residues and 20 monosaccharide residues; the latter are adjacent to (perhaps attached to, or buried in) the plasmalemma. By changing the sequence of the amino acid residues of the distal (N-terminal) part of the molecule, the antigenic stimulus presented to the host can be varied so that antibody elaborated against one variant does not cross-react with others (Fig. 4–2). Variation can occur about every three days, and a single trypanosome can produce perhaps 100 or more variable antigenic types (VATs); this is known because an apparently unlimited number can be produced *in vivo* by a clone of trypanosomes (i.e., a population derived from a single organism). The control of their production is not fully understood: presumably each glycoprotein is synthesized under the control of a single active gene, but how are the genes activated in turn? It was thought that host antibody might initiate the change, but as antigenic variation is now known to occur *in vitro*, without the presence of

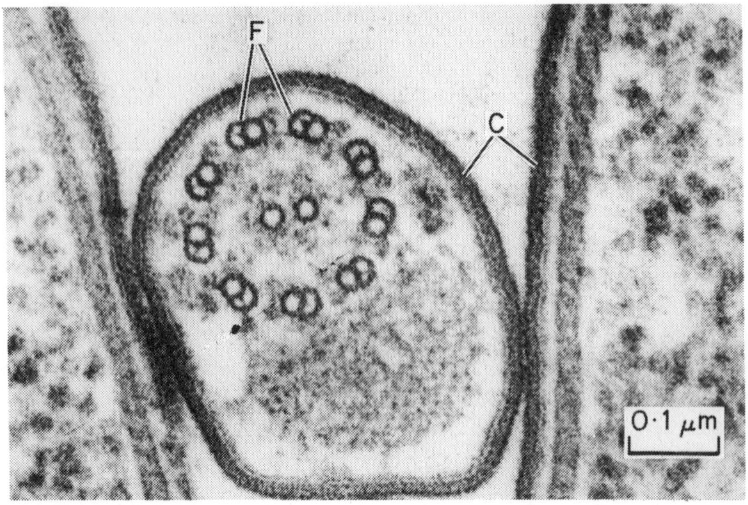

Fig. 4–1 Electron micrograph of section through flagellum and part of cell of haematozoic *Trypanosoma brucei*, showing glycoprotein surface coat (C) and flagellar microtubeles (F). (Micrograph by K. Wright, University of Toronto.)

antibody, activation appears to be an endogenous, possibly random, process with antibody playing a selective rather than inductive role (although with malaria parasites, antibody does seem to act as an inducer of antigenic change: see below). Furthermore, variation can occur early in an infection, before antibody could have been produced; it may occur even in the salivary glands of *Glossina*. Work involving 'cloning' trypanosomal DNA into a bacterial plasmid's genome suggests that re-arrangement of the sequence of genes on the trypanosomes' chromosomes may be related to changes in the antigen synthesized. Other recent work indicates that antigen synthesis may be controlled at the DNA-to-RNA transcription level. Antigenic variation results in the trypanosomes' always being one step ahead of the host's increasingly desperate secretion of antibodies, and thus avoiding complete elimination. Many parasites are, however, killed, and the infection tends to progress in a series of more or less regular waves of parasitaemia; thus another advantage to the parasite accrues – the infection is 'damped down' so that the host does not succumb too soon, thereby affording the trypanosomes a greater opportunity to infect *Glossina*. *T. brucei* and other salivarian trypanosomes also exert an immunosuppressive effect on the unfortunate hosts (see section 5.2.3), which reduces their ability to mount an effective immunological response to the trypanosomes themselves and to other antigens (such as bacteria); secondary infection is a common sequel of human trypanosomiasis.

Another consequence of antigenic variation is the extreme difficulty of producing a useful vaccine, although immunization against one particular variant can be achieved. However, trypanosome populations in one particular host mammal, and even in all individuals in certain geographical areas, tend to

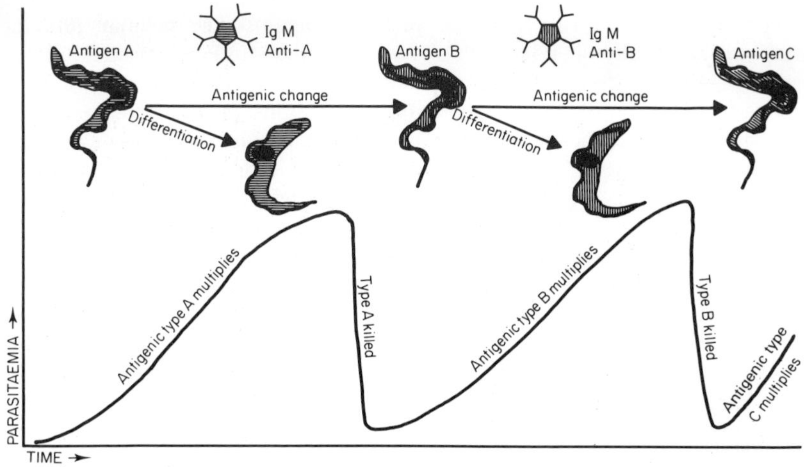

Fig. 4–2 Diagram illustrating successive production of different surface glycoprotein antigens by haematozoic *Trypanosoma brucei*. (Slightly modified from Baker, J. R., 1978, in *Companion to Microbiology*, Bull and Meadow, Longman, London.)

revert, after passage through *Glossina*, to one or a few basic antigens, and to give rise initially in their mammalian hosts to sequences including some identical (or at least cross-reacting) predominant antigens, so the outlook for vaccination may not be entirely hopeless.

Apart from the metacyclic trypomastigotes, the forms developing within *Glossina* lack the surface glycoprotein coat and hence express only the common, invariant antigens of the plasmalemma, which are screened, when in the vertebrate host, by the surface coat.

Antigenic variation has not been demonstrated amongst any trypanosomes other than those of the subgenus *Herpetosoma* (e.g. *T.lewisi*, mentioned above) and the salivarian subgenera *Trypanozoon, Nannomonas* and *Duttonella*. A similar process in *Plasmodium* and *Babesia* will be discussed in the next section.

4.2 Intracellular parasites

As mentioned above, *T. cruzi* probably evades its host's humoral immune response by inhabiting cells such as muscle cells, into which antibodies do not readily penetrate. To ensure transmission, *T. cruzi* trypomastigotes must periodically circulate in the blood, and during this phase they are certainly exposed to antibodies. Whether they survive by having changed antigenically since the previous haematozoic population appeared, or whether they merely rely on an extracellular life of limited duration while the host's B-cells are transforming into plasma cells before secreting antibody, is unknown. A strong cellular immune response develops against *T. cruzi*, and probably plays a more important role in removing trypanosomes from the blood than it does in *T. brucei* infections (in which phagocytosis seems to be mainly a 'follow-up' to lysis by antibody plus complement); human neutrophils plus antiserum ingest and kill *T. cruzi in vitro*, but whether this is true *in vivo* remains to be seen.

The intracellular forms may be protected against the cellular immune response also, although it has been shown that, in cell cultures, spleen cells from infected mice can destroy fibroblasts containing *T. cruzi*; again, the relevance of this to events occurring during natural infections *in vivo* remains to be ascertained. Paradoxically, some nosodemes (populations differentiated mainly by clinical characteristics) of *T. cruzi* inhabit cells of the mononuclear phagocytic series, which should be able to destroy them. Survival and replication within potentially destructive cells (macrophages) occur especially in species of *Leishmania* and *Toxoplasma*, the normal habitat of which, in their vertebrate hosts, is (or includes) cells of this type. The mechanisms by which they survive within these cells are not fully understood, and seem to differ between the different genera. *Toxoplasma* can inhibit fusion of lysosomes with phagosomes containing parasites; how it does this is not conclusively known, but parasitophorous vacuoles containing apparently healthy parasites are surrounded by a layer of mitochondria and endoplasmic reticulum which may act as a physical barrier to lysosomal fusion. *Leishmania*, in contrast, can survive even when bathed in lysosomal enzymes – provided the 'correct' combination of parasite and host species exists: e.g. *L. enriettii*, a parasite of guineapigs, survives in guineapig macrophages (even when they are 'activated'; see below) but not in those of mice, whereas *L. tropica*, the reservoir hosts of which are often rodents, is not killed by macrophages obtained from mice. *Trypanosoma cruzi* in macrophages probably survives by escaping from the phagosome into the host cell's cytoplasm. Non-specific activation of the macrophages (by prior injection of Bacillus Calmette-Guérin or BCG to the donor) increases their ability to kill *T. cruzi* and *Leishmania*, but not *Toxoplasma*. Pre-treatment of *Leishmania* with immune serum leads to their destruction without macrophages. Macrophages from specifically immune animals do not kill *Toxoplasma* but somehow inhibit its replication; 'immune' macrophages do however show enhanced ability to kill intracellular *Leishmania* and *T. cruzi*.

The other major group of intracellular protozoa is the Coccidia. The Piroplasmia induce production of circulating antibodies by the host and also undergo antigenic variation, but little is known of how these events relate to their survival (or death) in the host. Like the malaria parasites (see below), they are presumably vulnerable to humoral antibodies while in the process of transferring from one host cell to another, particularly when these cells are erythrocytes surrounded by plasma containing antibodies; probably it is at this stage that their ability to vary their surface antigens is of value. The malaria parasites (*Plasmodium* species) have been better studied. Initially, they inhabit host cells other than erythrocytes: in the species infecting mammals, these are liver parenchyma cells. Whilst in the liver cells, the parasites neither stimulate production of, nor are affected by, pre-existing humoral antibody: they are immunologically isolated from the host (though they react with antibody when exposed to it experimentally: for example, fixation and sectioning of an infected liver followed by treatment with fluorescein-conjugated immune serum results in fluorescence of the parasites when observed microscopically, illuminated by long wave-length blue-violet light). Once within erythrocytes, the parasites stimulate the host to produce antibodies. Malarial antigen is present on the surface of infected erythrocytes, and also is released into the plasma when the cells rupture at the end of each cycle of schizogony. Humoral antibody presumably reacts, therefore, not only with the liberated merozoites before they can enter other erythrocytes but also with the antigen-coated infected cells. There is some evidence that this coating of the erythrocytes with foreign (malarial) antigen leads to auto-antibody (anti-self) production and consequent destruction of unparasitized erythrocytes also, to the detriment of the host. The erythrocyte, therefore, does not provide a very safe environment for the malarial parasite, but the latter survives as, like the salivarian trypanosomes, it undergoes antigenic variation at approximately weekly intervals. Less is known of this phenomenon than of the corresponding behaviour of trypanosomes, but it seems that, in this instance, variation may be directly triggered by variant specific antibody, secreted by B-cells; the same or similar antibody (perhaps of higher affinity or in higher concentration?) can also be lethal. 'Helper' T-cells, whose response is not variant specific, expedite the B-cell response as the infection progresses and thus parasitaemias do not continually increase and the survival of the host is extended, to the benefit of the parasite. In fact, malaria parasites of man, except *P. falciparum*, seldom are, of themselves, lethal.

Less is known of the immunology of the eimeriine Coccidia. The intra-macrophagic pseudocyst of *Toxoplasma* has been discussed above. The tissue cyst of this parasite appears to be, like the liver stage of *Plasmodium*, immunologically isolated from the host though presumably the contained cystozoites are open to antibody attack when liberated from the cyst. The intestinal phase of *Toxoplasma* has been very little studied immunologically; it may well prove similar, in this as in other respects, to the better-known intestinal Coccidia such as *Eimeria*. Most species of *Eimeria* inhabit cells of the intestinal epithelium of the host, and are directly exposed to the host's antibodies only when entering such cells (as sporozoites) and transferring from one cell to another (as merozoites). There is no doubt that humoral antibodies are produced in response to eimerial infection in chickens, and there is no doubt that at least some degree of resistance to reinfection develops; what is doubtful is how this resistance is mediated. Generally, humoral antibody to *Eimeria* has not been found to have much protective effect, perhaps because it does not normally reach the intracellular parasites or the infective sporozoites; the latter probably enter either intermediate 'transport' host cells such as macrophages, or their final host cells, directly from the intestinal lumen without entering blood plasma. It has proved difficult also to demonstrate an effective cell-mediated component in resistance to *Eimeria*. Perhaps somewhat in desperation, it has been suggested that the observed protective immunity depends upon humoral antibody acting in conjunction with a hypothetical factor increasing permeability of localized areas of tissue or cells.

The development of means of surviving in spite of the host's immune response explains why no entirely effective and safe vaccine has been developed against a protozoal infection. However, as understanding of the processes involved increases, this desirable goal may yet be achieved.

5 Damaging the Host (Pathogenesis)

5.1 Invertebrate hosts

The most important pathogenic protozoan parasites of invertebrates are the Microsporida, a somewhat enigmatic group infecting predominantly insects. Some cause diseases of economic importance – e.g. pébrine of silkworm caterpillars (*Bombyx mori*), caused by *Nosema bombycis* and nosema disease of honey bees (*Apis mellifera*), caused by *N. apis*, among invertebrates; many insects, fish and some mammals, including occasionally man, are infected by other species (that infecting man is probably *Encephalitozoon cuniculi*). The damage done by Microsporida to their hosts results largely from cellular hypertrophy and destruction in the affected organs, resulting from schizogony of the intracellular parasites. The gregarine *Monocystis* infects the earthworm *Lumbricus*, which may suffer as a result of the destructive presence of many trophozoites within its spermatogonia. Apart from these and a few other little-studied associations, most parasitic protozoa (especially the dixenic forms) appear to be harmless to their invertebrate hosts. The only known example of a parasite of medical or veterinary importance which harms its vector is *Trypanosoma rangeli*. This organism infects human beings and other primates in South America, but produces no apparent disease. It is transmitted (like the pathogen *T. cruzi*; see section 5.2) by Reduviidae (bugs), which it damages by depleting their haemolymph of certain essential amino acids; infected bugs have significantly reduced survival. In contrast, even swarming infections of the midgut of *Glossina* with trypanosomes do the insects no apparent harm. Experimental infections of mosquitoes with malaria resulting in almost complete coverage of the midgut with oocysts seem equally harmless, as does the virtual blockage of a sandfly's pharynx with leishmanial promastigotes. This harmlessness benefits the parasite, since vector longevity increases the likelihood of its reaching the next host.

5.2 Vertebrate hosts

The statement with which the previous section ends is equally true of the vertebrate host, early death of which is likely to result in early death of the parasite population also. Thus, generally, avoiding excessive damage to the host is as important to the parasite as averting its own destruction by the host. A delicate balance must be struck and, in general, the degree to which this has been achieved reflects the 'success' of the parasite and the evolutionary duration of its association with the host. It is a truism, but probably usually true, to state that pathogenic associations are recent ones. It is often forgotten that most parasitic associations are more or less in balance: only the exceptions impose themselves on our consciousness. The several species of amoebae and flagellates living harmlessly within our intestines are of no great import; even less so are the many species of Haemosporina, *Trypanosoma*, Eimeriina, Piroplasmia and others infecting the world's fauna (and occasionally flora: e.g. the kinetoplastid flagellate *Phytomonas* in Euphorbiaceae) with no obvious harmful effect. It is economically sensible to concentrate scant resources on parasites destructive of human welfare, either directly by causing illness or indirectly by reducing the available supply of protein, but it is also sensible to preserve a biological perspective by remaining aware of the nine-tenths of the parasitological iceberg, composed of harmless (and occasionally beneficial) associations, submerged beneath the level of detectable pathogenicity.

The many ways in which parasites, protozoa and others, damage their hosts can be considered in three major groups: direct, local damage; direct, diffuse damage; and indirect damage. These terms include, respectively, mechanical damage produced by the parasite in its immediate environment; damage produced directly, but at a distance, by secretions (toxins) of the parasite; and damage resulting from interaction (usually immunological) between the parasite and its host. It is not always possible to distinguish sharply between these types, but the attempt can be helpful.

5.2.1 Direct, local damage

This kind of effect includes the mechanical destruction of erythrocytes by rupturing schizonts of *Plasmodium* or *Babesia*, and of cardiac muscle cells by *Trypanosoma cruzi*. Also within this category is the damage to duodenal mucosae resulting from the attachment of *Giardia muris*. This organism uses some of its four pairs of flagella to induce flow of the surrounding liquid medium, thus reducing the hydrostatic pressure beneath its ventral holdfast ('sucking disc') enough to hold the parasite against the substratum by dynamic suction. The result is a lesion on the mucosal cell surface which looks, in the scanning electron microscope, as though made by pressure from the sharp end of a cork borer (Fig. 5–1). Strangely, it seems that *G. muris* is not harmful to its rodent hosts, while the human pathogen *G. lamblia* (Fig. 5–2) does not, apparently, cause these 'cork-borer' lesions. This suggests that the direct, mechanical damage to the host is not very important. Perhaps this is true of many direct effects: there is little evidence that the uncomplicated anaemia resulting solely from mechanical destruction of erythrocytes by *Plasmodium* harms the host significantly (see section 5.2.3).

Examples of more severe direct damage include that done by eimeriine coccidia, particularly those living deep in the intestinal wall such as second generation schizonts of *Eimeria necatrix* in domestic fowls; when these mature and rupture their host cells, the resulting damage may lead to considerable bleeding. Also of this type, probably, is the usually fatal disruption of brain architecture by the facultatively parasitic amoeba *Naegleria fowleri*, which may enter the brain (fortunately rarely) via the olfactory tract. The intestinal

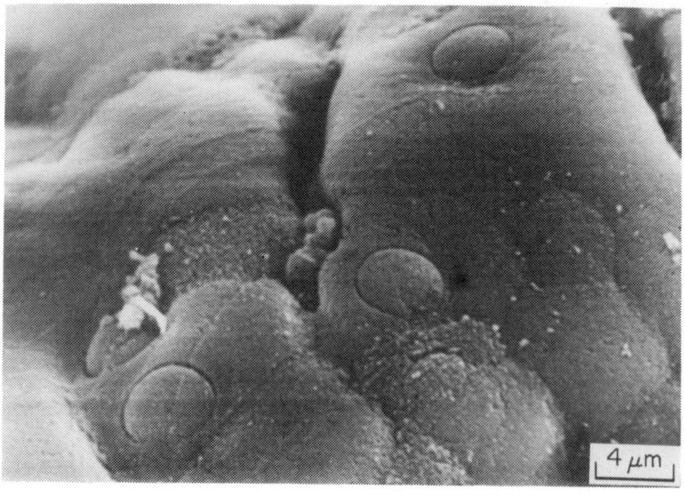

Fig. 5–1 Circular lesions on mucosa of small intestine of mouse, resulting from attachment of *Giardia muris* (scanning electron micrograph by D. V. Holberton, University of Hull).

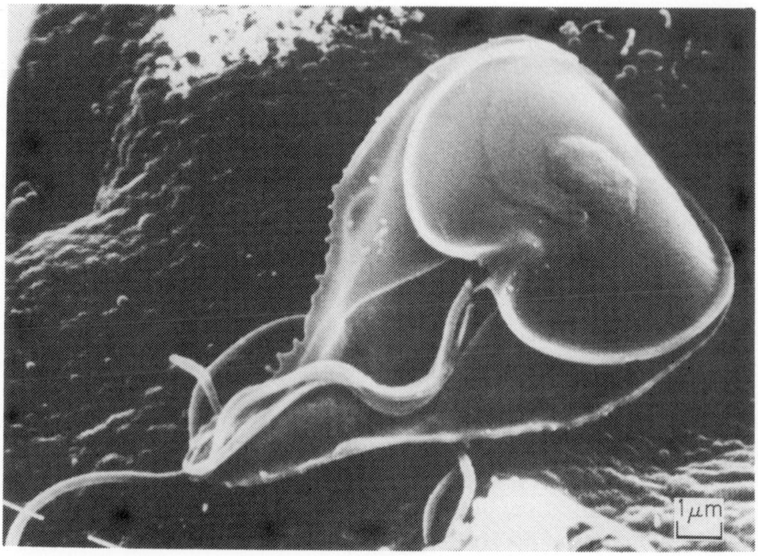

Fig. 5–2 Scanning electron micrograph of *Giardia lamblia* in biopsy of human jejunum (micrograph by K. Vickerman, University of Glasgow).

amoeba *Entamoeba histolytica* probably also damages its host, when it does so, by direct damage. More often than not, this obligate parasite lives harmlessly in the lumen of the large intestine, feeding on food debris and bacteria. Sometimes, for reasons not yet understood but possibly involving the composition of the host's diet, the amoeba penetrates the gut wall causing ulceration, bleeding, and dysentery. From these ulcers, the parasites may be carried in the blood stream to the liver, lung, brain or other organs; there they may remain, multiply and feed by phagotrophy (section 3.1) on tissue cells including erythrocytes from eroded capillaries. The resulting lesions are misnamed 'abcesses', as they contain no bacteria. The amoebae may secrete an enzyme ('cytolysin') to assist the feeding process, but this is uncertain. The damaged tissue seems to release substances inducing fever ('pyrinogens') into the circulation, but apart from this there is no diffused pathogenic effect.

Direct mechanical damage to the liver by *Leishmania donovani* may result from parasitization and proliferation of the Kupffer cells, which interferes with absorption from the sinusoids as well as disarranging the liver structure, but an indirect effect is important also in this disease (section 5.2.3).

5.2.2 Direct, diffuse damage

Although toxin production by parasitic protozoa is frequently proposed as a pathogenic mechanism (usually when no other can be identified), accredited examples are rare. A toxin has been suggested as the agent responsible for neurone destruction by *Trypanosoma cruzi* but, as usual, direct evidence is lacking and other possibilities exist, such as cross-reacting or auto-antibodies. Lipopolysaccharide (LPS) may be produced by *Plasmodium berghei* and *Trypanosoma cruzi*, and infection of mice with *P. berghei, P. vinckei, Babesia microti* or *B. rodhaini* increases their sensitivity to shock induced by LPS. It has therefore been proposed that excretion of LPS from malaria parasites may induce shock, fever, and tissue damage. The cyclical release of LPS or other toxic material, resulting from the synchronization of cycles of erythrocytic schizogony (see section 6.2), is probably responsible for the characteristic cyclical recurrences of fever every 2 or 3 days in human malaria. LPS may also produce blackwater fever (acute haemolysis and kidney failure, revealed by excretion of altered haemoglobin in the urine – hence 'blackwater') – a rare, but serious, complication of *P. falciparum* infection of man, often associated with quinine therapy. An alternative suggestion for the genesis of blackwater fever, supported by more experimental evidence, is that quinine counteracts the normal inhibition by plasma proteins of erythrolysis induced by free fatty acids (FFA) in the plasma. This effect may be exacerbated by reduction in the plasma protein concentration during malaria and excretion of FFA by the parasites. FFA is also a likely candidate for the toxin proposed to account for the tissue damage in malaria resulting from the cells' inability to use available oxygen (histotoxic anoxia). Such damage could account, at least in part, for much of the pathology of acute *P. falciparum* malaria – including liver malfunction, kidney failure and brain haemorrhage (see also section 5.2.3).

5.2.3 Indirect damage

This section could be headed 'immunopathology'. During the last few years, it has become increasingly recognized that parasitic disease, as distinct from infection, may result to a considerable extent from the host's misguided defensive efforts, as a result of either auto-antibody production or tissue damage due to formation of substances called kinins. Immune complexes (antigen + antibody + complement) adsorb and activate a substance in the blood called Hageman factor, which in turn starts a series of 'cascade' reactions resulting in production of plasmin, fibrin and kinin. Kinin is one of a group of low molecular weight peptides, which increase vascular permeability leading to leakage of plasma into the surrounding tissue. The consequent oedema disrupts the architecture of the tissue. Kinin release may be especially important in infections with parasites practising antigenic variation, which therefore constantly stimulate the host to produce new antibodies to the changed antigen. Thus the ensuing peptide production and tissue damage may be extended both in time and space.

Auto-antibody production has been invoked to explain the destruction of uninfected as well as parasitized erythrocytes during malarial infection, but other explanations exist (section 5.2.2); more than one process may of course be involved. Circulating immune complexes may contribute to the anaemia of malaria and African trypanosomiasis, because their attachment to the erythrocyte surface would encourage phagocytes to engulf the affected blood cells. Recently an antibody has been reported in the serum of many patients with chronic Chagas's disease which reacts with the patient's own endocardium, vascular endothelium and muscular interstitial tissue (EVI antibody). EVI antibody appears not to be strictly an auto-antibody but an antiparasite (*T. cruzi*) antibody cross-reacting with host tissues.

Kinin release following antigen-antibody interaction in the blood almost certainly contributes to tissue damage in African trypanosomiasis and malaria. It may account, at least partly, for the brain damage characteristic of both *P. falciparum* malaria (haemorrhages) and *T. brucei* trypanosomiasis (meningoencephalitis), and the more general chronic inflammation seen in these, and Chagas's disease.

In cutaneous leishmaniasis, the course of the disease is largely determined by the balance between the host's immune response (or non-response) and the parasite itself. Some strains or species of *Leishmania* (see section 3.2.3) have highly immunogenic antigens, leading (if the host's immune response is good) to strong delayed hypersensitivity phenomena and limitation of the infection to a single skin lesion. Others have antigens which are more tolerogenic; this (especially if the host's immune response is poor) may result in the spread of parasitized macrophages throughout large areas of skin – a most disabling, unsightly and refractory condition called diffuse cutaneous leishmaniasis. A spectrum of disease between these two extremes exists, depending on the resultant of the two separate components – host's and parasite's. In leishmaniasis, the important defensive response of the host is cell-mediated. Circulating antibody may be present, but has little effect except perhaps a negative one by blocking the attack of potentially cytotoxic lymphocytes. Cutaneous leishmaniasis is rarely if ever fatal. In visceral leishmaniasis, the precise cause of death is unknown. Tissue damage, especially to the liver, may well result from antigen-antibody reactions, and may contribute to the anaemia and wasting which occur; there is also a direct effect (section 5.2.1).

A third means of indirect damage to the host is immunodepression, known to occur during infection with *Plasmodium, T. cruzi* and subspecies of *Trypanosoma brucei* in human beings and experimental animals, and in the latter if experimentally infected with *Babesia, Toxoplasma* or *Leishmania*. It has been most studied in malaria and trypanosomiasis. The underlying mechanism is unknown. Suggestions include interference with antigen processing by macrophages overburdened with parasite debris; an upset in cooperation between T- and B-lymphocytes; polyclonal exhaustion of lymphocytes following secretion of a non-specific T- or B-cell mitogen (possible a saturated fatty acid); antigenic competition; or the generation of unidentified suppressor cells. None of these ideas seems to explain all the observed events; probably more than one mechanism is involved, even in the response of a single host species to a single parasite species. In malaria and babesiasis, the depression seems to involve mainly B-lymphocytes, while in African trypanosomiasis both categories (T and B) are concerned. It is difficult to imagine an advantage accruing to the parasite from immunodepression, unless the reduced responsiveness relates also to the inducing parasite itself. There is some evidence that this is so: antibodies against antigenic variants of *Plasmodium* and *Trypanosoma*, produced late in the infection of experimental animals, have lower affinity for the parasites than do those synthesized earlier. Perhaps this mechanism plays a part in maintaining the delicate balance between parasite survival and host survival, referred to in section 5.2. On the other side of the scale, in this balance, is the fact that the later antibodies in experimental malaria infections appear to be synthesized more quickly after the appearance of each antigenic variant, perhaps as a result of the building up of a population of 'helper' T cells which recognize invariant antigens in the parasite. It is easier to see how immunodepression harms the host; death from intercurrent bacterial infections such as pneumonia is common in untreated sufferers from African trypanosomiasis, and the virus-induced cancer Burkitt's lymphoma is more prevalent in areas of Africa where malaria is very common, probably because malaria depresses the patient's ability to mount an effective immune response to the carcinogenic virus.

6 Leaving the Host

6.1 One-host parasites: development outside the host

Leaving the host is seldom a problem to monoxenic parasites. Most emerge in the host's faeces, the few exceptions being those with bizarre habitats such as the mouth (*Trichomonas tenax, Entamoeba gingivalis*), genital cavities (*Trichomonas vaginalis*) or skin of the genitalia (*Trypanosoma equiperdum*); for these the exit route is obvious (section 2.1.2). Parasites in the intestinal lumen or wall (*Entamoeba histolytica* and most eimeriine coccidia, for example) have the shortest route to the outside. For others, such as *Eimeria steidai* in rabbits' bile duct epithelium, the route is more tortuous.

A necessary preliminary for survival in the hostile outer world for all protozoa of this group except *E. gingivalis*, trichomonads, and *Trypanosoma equiperdum*, is encystment. Those species that do not form cysts presumably often rely on more rapid direct transmission by kissing or coitus to enable them to survive without this protection, though this cannot apply to *Pentatrichomonas hominis* of the human gut. One zoomastigophorean, *Histomonas meleagridis*, lives in the liver and caecum of chickens and other gallinaceous birds, causing the disease 'blackhead' in turkeys. *H. meleagridis* has no cyst, and has overcome the problem of protection during transmission by travelling as a passenger within eggs of a parasitic nematode, *Heterakis gallinae*. *Dientamoeba fragilis*, a related parasite living commensally in human intestines, may also 'ride' in the eggs of some nematode, but this has never been proved.

Little is known of stimuli inducing encystment. Sometimes, *in vitro*, reducing the available food supply encourages encystment of *Entamoeba* species. *E. histolytica* never encysts when living within the intestinal wall, liver or other organs. In these situations, presumably, nutrients abound and encystment would be biologically pointless since amoebae in such habitats have no access to the outside world. Paradoxically, the occasionally parasitic amoeba *Naegleria* does encyst in mammalian tissues, but it is only an accidental parasite probably never transmitted directly from mammal to mammal. Encysted amoebae and flagellates usually undergo one to three nuclear divisions, but thereafter remain quiescent until the excystment stimuli are experienced (section 2.1.1).

Encystment of eimeriine coccidia is presumably stimulated by fertilization of the female gametocyte. Vacuoles containing the ingredients of the future oocyst wall ('wall-forming bodies') have been identified in the female gametocyte and zygote of many species. Encystment begins, and may be completed, before the zygote leaves its host cell, which it does presumably by mere mechanical disintegration of the dead or dying cell. More activity occurs within eimeriine oocysts than within amoebic or flagellate cysts, as sporogony occurs therein. This may or may not be completed before the oocyst is voided by the host. After its completion, no further activity occurs until the oocyst is exposed to appropriate hatching stimuli (section 2.1.1).

At least one group of parasitic Protozoa has evolved an interesting means of ensuring that cysts are produced at a time of year when young, susceptible new hosts are available. These are the opalinids, usually nowadays placed in the Sarcomastigophora (Table 1). They are parasites of the rectum of frogs, transmitted by a cyst in the adult frog's faeces. The feeding habits of adult frogs make it unlikely that they would be exposed to infection; thus a mechanism ensuring that cysts are produced only when tadpoles are available benefits the parasite, by concentrating its transmissive effort when most likely to succeed. Opalinids therefore encyst only during the breeding season of the frogs, apparently stimulated by gonadotrophic hormones from the frog's pituitary gland. Other examples of timing production or availability of transmissive stages are described in section 6.2.

6.2 Two-host parasites: escaping from the vertebrate host

Most dixenic parasitic protozoa leave their vertebrate hosts via an insect vector. The few exceptions include such organisms as *Toxoplasma gondii* which, in its encysted exoenteric stage in various warm-blooded vertebrates (secondary hosts), can effectually leave its host only if the latter is eaten by a predatory primary host (section 3.2.2). This mechanism is important in the epidemiology of human toxoplasmosis, since, together with the swallowing of oocysts in cats' faeces and diaplacental infection, it forms one of the commoner means of human infection (man being a secondary host of *T. gondii*). The exoenteric ('tissue') cysts may be eaten in raw or undercooked meat (steak tartare or rare hamburger, for example). A small epidemic among North American university students, referred to in section 3.2.2, was traced to the consumption of too hurriedly prepared hamburgers by students rushing to a lecture by a famous speaker. Cysts can also be ingested from the fingers of butchers or cooks handling raw meat (including rabbits). Although very thin-walled, the tissue cysts of *T. gondii* are mechanically tough and resistant to gastric juice. This route of transmission is undoubtedly important also in infecting the primary host, in which the enteric phase occurs. The primary host is invariably a carnivorous mammal (e.g., *Felis domesticus* for *T. gondii*) or bird (*Buteo buteo* for *Frenkelia buteonis*, of which the secondary host is the bank vole *Clethrionomys glareolus*).

Adaptations for infection of an invertebrate vector are legion. Presence in the circulating blood of the vertebrate host is almost *sine qua non*, but *Leishmania tropica* and its close relatives, which are often restricted to a single lesion on the skin of the vertebrate host (section 5.2.3), rely on their vector (a sandfly – Phlebotominae) acquiring parasites by plunging its proboscis through the infected region. Malaria parasites have specialized sexual stages (gametocytes) destined to infect the vector (a mosquito – Culicidae), which develop from certain of the merozoites produced by erythrocytic schizogony. The reason(s) why some merozoites become asexual, and others sexual, individuals is (are) unknown. Erythrocytic schizogony of malaria parasites is usually closely synchronized, occurring at about the same hour of the day in many strains. The duration of the schizogony cycle is an inherent, presumably genetic, characteristic of the parasites, but it is entrained, and its astonishing accuracy maintained, by external stimuli derived from the host's diurnal cycle of activity. The reason for this piece of protozoan time-keeping seems to be related to the advantage conferred on the parasite by having successive broods of mature gametocytes (derived from the synchronously produced merozoites) available at the time of day when the vector species is most likely to feed. With human malaria parasites, all of which are transmitted by anopheline mosquitoes, this is usually during the evening and early night. In at least some species of *Plasmodium* the duration of fertility of gametocytes is limited to perhaps a few hours. It is therefore biologically important to the parasite that these sex-crazed gametocytes should be available to the right species of mosquito, in which alone they can complete their natural development and thus ensure the transmission, and therefore survival, of the parasite. Unfortunately this beautifully reasoned argument crumbles somewhat due to the fact that one species of *Plasmodium* which infects man (*P. falciparum*), while still having a synchronized 48 hour schizogony cycle, apparently produces gametocytes which remain sexually mature for several days or even weeks. This paradox is as yet unresolved.

Trypanosoma brucei and probably *T. cruzi* have certain forms (morphs) preadapted to infection of the vector. Other species apparently do not, but some (*T. congolense* in experimentally infected rats, *T. minasense* in *Callithrix* marmosets, and *T. rotatorium* in *Rana clamitans*) have daily cycles of abundance in their host's peripheral blood which may be related to the vectors' feeding activity, though this has yet to be demonstrated. Peak parasitaemia of *T. congolense* occurs at night, which does not appear to fit with the vector's (*Glossina* spp.) daytime activity. However, the rat is an unnatural host of *T. congolense* and is nocturnally active, whereas many of this trypanosome's natural hosts (ungulates) are active during daylight. If, therefore, the prevalence of trypanosomes in the peripheral blood is controlled by the host's activity (as the timing of malarial schizogony appears to be), the greatest numbers might occur during the day in natural hosts, at the time when vectors are more active. However, this is at present speculative. Peak occurrence in peripheral blood occurs in the afternoon with *T. minasense*, and around noon with *T. rotatorium;* the vectors of these two species are unknown. When less common peripherally, the trypanosomes appear to congregate in the visceral circulation – of the kidneys, with *T. rotatorium*. The mechanism controlling this is not known in full. With *T. rotatorium*, the primary control seems to be the environmental light-dark cycle, with environmental temperature playing a subsidiary role. How the light-dark effect is perceived is also a puzzle: apparently neither the eyes nor the pineal body of the frogs is involved.

While it is easy to understand the advantage to the parasite of being abundant in the peripheral blood when vectors are likely to feed, it is less easy to understand what they gain from 'hiding' in the deeper circulation at other times. Possibly they are partially protected from host antibodies or phagocytes. Some microfilariae behave very similarly (see LYONS, 1978).

7 Evolution

The evolution of parasitic protozoa is a subject on which much has been written but little is known. Clearly it is a subject on which one can only speculate, there being no fossil record to confirm or refute the products of one's imagination. What follows is a brief summary of current concepts, and should be accepted (if at all) only in the speculative spirit in which it is offered (see Fig. 7–1).

The only certainty in this subject is that parasitic forms must have evolved from free-living ones. Popular and likely candidates for the hypothetical basal protozoan (and algal) stock are the relatively unspecialized amoeboflagellates (Sarcomastigophora). From these the amoebae (Sarcodina) and flagellates (Mastigophora) may have evolved by loss of some of their primal pluripotentiality – that is, their ability to move by means of both pseudopodia and flagella. These powers have not however been entirely lost by any major group of animals, let alone protozoa. Flagella have been retained by the male gametes of man and malaria parasite alike, as has amoeboidicity of certain cells at least; mammalian macrophages and *Plasmodium vivax* trophozoites are examples. Many green plants have also retained these features at certain stages of their life cycles.

Amongst the amoebae, parasitism has probably been adopted independently by various groups without ever becoming the possession of any major group. The intestinal amoebae are probably heterogeneous and of polyphyletic origin, with only one genus (*Entamoeba*) sufficiently interested in the parasitic way of life to have diversified very far in that direction. Even *Entamoeba* has produced only a limited number of species (four in man – *E. histolytica, E. coli, E. hartmanni, E. gingivalis* – and several elsewhere). The so-called 'pathogenic soil amoebae' (predominantly *Naegleria*) are facultative parasites which may or may not be en route to a more completely parasitic life. Flagellates too have probably adopted a parasitic life style on various isolated evolutionary occasions; those inhabiting the intestines of vertebrates are a heterogeneous group. Parasitic specialization among the flagellates has been indulged in pre-eminently by the obligate symbionts (Hypermastigida) of termites (Isoptera) and cockroaches (Insecta, Orthoptera, Blattidae). These flagellates are essential for the wood-eating arthropods as they provide their only source of cellulase. Another group of flagellates specializing in parasitism, and of more direct concern to ourselves, is the Kinetoplastida. Within this group evolution probably progressed from monoxenic promastigotes parasitic in the gut of an early invertebrate group ancestral to molluscs, annelids and insects, through a variety of forms dwelling in the guts of insects to dixenic parasites of insects and vertebrates (*Leishmania, Trypanosoma*). Kinetoplastida are essentially parasites of insects, and only secondarily associated with vertebrates. However, the ultimate trend (so far) in their evolution is emancipation from the primal invertebrate host, exemplified by certain trypanosomes (*T. evansi, T. equinum* and *T. vivax viennei*) which have dispensed with cyclical development therein and, finally, *T. equiperdum*, which

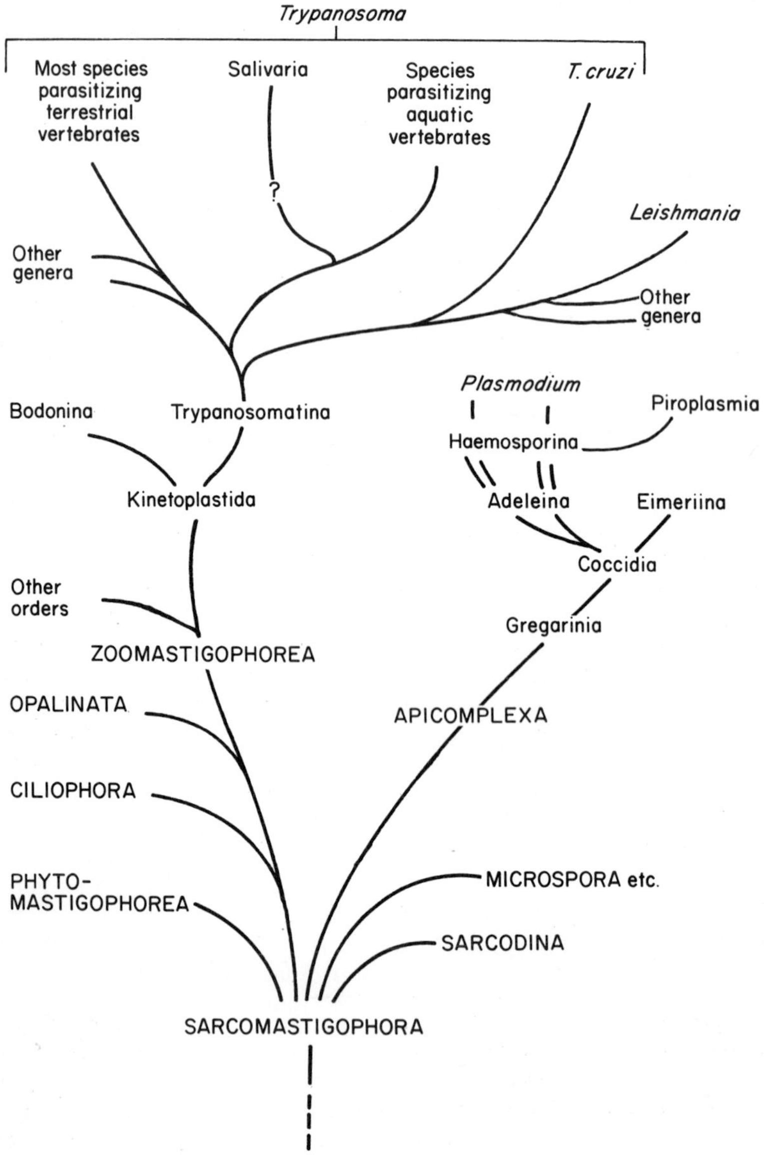

Fig. 7–1 Speculative 'family tree' (evolutionary dendrogram) of the protozoa. The query mark on the line leading to 'Salivaria' indicates their possibly relatively recent origin from amphibious reptiles in tropical Africa (see BAKER, 1974). 'Microspora etc.' includes the minor protozoan phyla Microspora, Ascetospora and Myxozoa. (Modified from BAKER, 1973.)

has dispensed entirely with the need of an insect vector by substituting a horse's penis. In some parts of South America, *T. equinum* has replaced the original insect vector by a mammalian one, the vampire bat *Desmodus rotundus*. The origin of the somewhat aberrant group of trypanosomes transmitted by tsetse flies (*Glossina*) has been the object of some discussion; probably they arose fairly recently from among the trypanosomes of reptiles transmitted by leeches (Annelida, Hirudinea). Adaptive radiation among this group of trypanosomes (the salivarian trypanosomes) seems to be in active progress at present, especially within the subgenus *Trypanozoon*, where the bounds of species, subspecies, demes, etc. are as firm as the shifting desert sands. The taxon of indeterminate rank known as '*rhodesiense*' is probably evolving even now, before our very eyes, from *T. brucei*.

The major group of protozoa which has specialized par excellence in parasitism is the Apicomplexa; all its members are parasites. Probably they originated as monoxenic parasites in the intestines or other body cavities of invertebrates like many present-day gregarines (order Gregarinida) such as *Monocystis* (in earthworms). From such habitats the apicomplexans found their way into similar locations within the evolving vertebrates, probably by being accidentally swallowed by the latter (e.g., Eimeriina). Next, they became dixenic either by adopting secondary development within another, facultative vertebrate host species (e.g., *Toxoplasma* and its allies) or by colonizing the vertebrate's blood cells and thus gaining access to blood-feeding insects and other arthropods (Acarina). This last development seems to have proceeded along two main lines – those of the adeleine haemogregarines and the Haemosporina respectively. The Piroplasmia probably represent an offshoot of one or other of these lines, most likely the latter. Within the Haemosporina, it is thought that the genera which do not undergo schizogony within erythrocytes are the more primitive (e.g., *Haemoproteus, Hepatocystis, Leucocytozoon*). The introduction of the additional erythrocytic schizogony cycle was probably a relatively late experiment in increasing the parasites' replicative ability: it gave rise to the genus *Plasmodium*.

Another group of entirely parasitic forms is the Microspora. The evolutionary origins of this group are obscure, but their closest living relatives may be the Apicomplexa. They form a relatively distinct and homogeneous group.

Finally, the ciliates. Probably the acme of protozoal evolution, this group has in general spurned parasitism. A few members of a few disparate groups have independently become obligate endoparasites (e.g., *Balantidium*) and within the genus *Tetrahymena* are species ranging from entirely free-living, through occasionally, facultatively parasitic forms to those with almost completely parasitic life cycles. Other ciliates have taken up ectoparasitism, mainly on fish (e.g., some Suctoria and *Trichodina* species). The only group to specialize in parasitism are the ciliate commensals in the rumen of Bovidae and, to a lesser extent, the caecum of Equidae; a good example is the ciliate order Entodiniomorphida. These are a homogeneous group with some remarkable physiological and morphological specializations; they deserve a book to themselves, as do the termite flagellates.

Appendix:

summarized details of parasitic protozoa of medical or veterinary importance (based mainly on BAKER, 1973).

Some species of *Eimeria* and *Isospora* which infect domestic animals and man.

Host	*Species*	*Habitat*	*Pathogenicity*
Man	*I. belli*	Upper small intestine	Mild-severe
Cattle	*E. bovis*	Small and large intestine	Fairly high
	E. zurnii (+17 other species)	Small and large intestine	Fairly high
Sheep and goats	*E. ahsata*	?	Moderate
	E. arloingi (+8 other species)	Small intestine	Moderate
Pigs	*E. debliecki* (+5 other species)	Small and large intestine	Mild-moderate
	I. suis	Small intestine	Mild
Chickens	*E. necatrix*	Small intestine and caeca	High
	E. tenella (+6 other species)	Caeca	High
Turkeys	*E. adenoeides*	Small and large intestine	Moderate
	E. meleagrimitis (+5 other species)	Small intestine	Moderate
Ducks	*E. bucephalae* (+1 other species[1])	Small intestine	High
Geese	*E. truncata* (+5 other species)	Kidney tubules	High
Rabbits	*E. stiedai*	Bile duct of liver	High
	E. irresidua	Small intestine	Moderate
	E. magna (+2 or more other species)	Small intestine	Moderate
Dogs and cats	*I. canis* (dog only)[2]	Small intestine	Moderate
	I. felis[2] (+1 or 2 other species of *Eimeria* in cats only, and 1 other species of *Isospora*)	Small intestine (rarely large)	Moderate

[1] A species of another genus, *Tyzzeria perniciosa*, is pathogenic to ducks.
[2] Recently transferred to *Levineia* gen. nov. by some workers.

		Trophozoite	Cyst				
		Size (μm)[1]	Size (μm)[1]	No. of nuclei (mature)	Chromatoid bodies	Special features	Chemotherapy
Entamoeba histolytica	Large intestinal lumen and wall; also other tissues	10–40	9.5–15.5 (~12)	4	Broad, blunt ends	—	Metronidazole
E. hartmanni		9–14	4–10.5 (~7.4)	4		—	—
E. coli		15–30	10–30 (~17)	8	Thin, sharp ends	—	—
Endolimax nana	Large intestinal lumen only	6–12 (~8)	6–9 × 5–7	4	None	—	—
Iodamoeba buetschlii		5–20 (~11)	9–15	1	None	Persistent glycogen vacuole[2]	—
Naegleria fowleri	Brain[3]	7–12 (~10)	7–15	1	None	Spherical	Amphotericin B[4]
Acanthamoeba spp.	Various tissues[3]	8–40	8–29	1	None	Polyhedral or biconvex	Sulphadiazine

[1] Means in parentheses.
[2] Stains golden-brown colour with iodine solution.
[3] Also free-living in soil and water; *N. fowleri* has a biflagellate free-living phase.
[4] Not very effective.

Parasitic flagellates inhabiting the human intestine.

	Trophozoite			*Cyst*		
Species	*Size* (μm)	*No. of flagella*	*Other characters*	*Size*	*No. of nuclei*	*Other characters*
Chilomastix mesnili	6–24 × 3–10	3 + 1[1]	Cytostome	7–10	1	Lemon-shaped; cytostome fibril
Enteromonas hominis	4–10 × 3–6	3+1	No cytostome	6–8 × 4–5	4	None
Retortamonas intestinalis	4–9 × 3–4	1 + 1	Cytostome	4.5–6	1	As *C. mesnili*
Giardia lamblia[2]	10–20 × 5–10	4 pairs	'Sucker'; two nuclei; median bodies	8–14 × 6–10	4	Fibrils and axonemes
Pentatrichomonas hominis	8–20 × 3–14	5 + 1	Undulating membrane; axo-style; cytostome	No cyst		
Dientamoeba fragilis	7–12	None[3]	Often binucleate[3]	No cyst		

[1] This notation implies that there are three free anterior flagella and one recurrent one.

[2] Pathogenic sometimes; treated with metronidazole.

[3] Superficially resembling an amoeba, with which it used to be classified, *D. fragilis* is now recognized as an aberrant trichomonad flagellate. It has a prolonged telophase, so many trophozoites are binucleate.

Main morphological features of the haematozoic trypomastigotes of some mammalian trypanosomes.

Subgenus	Species	Average length (μm)	Average breadth (μm)	Kinetoplast Size	Kinetoplast Position	Posterior end	Free flagellum
Schizotrypanum	*T. cruzi*	15–24	1.5	Large	Sub-terminal	Pointed (short)	Long
Duttonella	*T. vivax vivax* } *T. v. viennei* }	21–26	1.5–2 }	Large	Terminal	Blunt, rounded	Long
	T. uniforme	14–17	1.5 }				
Nannomonas	*T. congolense*	12–18	1–2 }	Medium	Sub-terminal, marginal	Usually blunt	Usually absent
	T. simiae	17–18	1–2 }				
Pycnomonas	*T. suis*	15	3.5	Small	Sub-terminal	Pointed (very short)	Short
Trypanozoon	*T. brucei*						
	Slender forms	30	1.5 }	Small	Sub-terminal	Blunt, often truncate	Long
	Stumpy forms	18	3.5 }			Blunt, rounded	Usually absent
	T. evansi } *T. equiperdum* }	20–28	1.5	Small	Sub-terminal	Blunt or truncate	Long
	T. equinum	20–28	1.5	Absent	—	Blunt or truncate	Long

Trypanosoma species of medical and veterinary importance.

Subgenus	Species	Geographical distribution	Main vertebrate hosts Wild	Domestic
Schizotrypanum	*T. cruzi*	S. and Central America	Armadillos, rats	Dogs, cats, MAN
Duttonella	*T. v. vivax*	W., Central and E. Africa	Waterbuck, reedbuck, eland, giraffe, bushpig	Cattle, sheep, goats, horses, donkeys (often pathogenic)
	T. v. viennei	Mauritius, W. Indies, S. America	?	As for *T. v. viv*
	T. uniforme	W. Uganda, E. Congo, (Tanzania and Zululand rarely)	Antelope	Cattle, pigs, (rarely if ever pathogenic)
Nannomonas	*T. congolense*	W., Central and E. Africa	Many antelope, giraffe, eland, wildebeeste	Cattle, sheep, h donkeys, pigs, rarely camels (c pathogenic)
	T. simiae	W., Central, E. and parts of S. Africa	Warthog	Pigs (very path
Pycnomonas	*T. suis*	Tanzania, Rwanda-Burundi, ? Congo	Warthog	Pigs (pathogen to young)
Trypanozoon	*T. brucei*	W., Central, E. and parts of S. Africa	Many antelope, warthog, wildebeeste	Camels, horses donkeys, dogs cattle, pigs, she goats (may be genic but more chronic); MAN
	*T. evansi**	N. Africa, Asia, S. China, Philippines, Mauritius, Central and S. America (eliminated from N. America and Australia)	Tapir (in America), deer (in Mauritius), vampire bat (S. America)	Camels, horses, donkeys, mules pathogenic); ca water buffalo, s goats, dogs, Ind elephant
	*T. equiperdum**	S. America, USSR, Iran, ? N. and S. Africa	None	Equines (patho
	*T. equinum**	S. America	Vampire bat and capibara	Horses (pathog donkeys, mules pathogenic); ca sheep, goats, pi (chronic)

*Taxonomic status uncertain; *T. equinum* lacks a kinetoplast visible by light microscopy; so do some

	Transmission		Drug treatment	
tory	*Cyclical*	*Non-cyclical*	*Prophylactic*	*Curative*
s	Reduviidae	—	None	Nifurtimox; benznidazole (toxic and not very effective)
lly none ith supplement)	*Glossina* spp.	Biting Diptera		
lly none	None	Biting Diptera		
lly none	*Glossina* spp.	Biting Diptera		
:s (not all are e; enicity	*Glossina* spp.	Biting Diptera (less important)	Isometamidium chloride; quinapyramine prophylactic; pyrithidium bromide	Isometamidium chloride; quinapyramine sulphate; pyrithidium bromide; diminazene aceturate; homidium bromide
ithecus ys (very enic) and :tomized	*Glossina* spp.	Biting Diptera (less important)		
known	*Glossina* spp.	?	?	?
s, rabbits, ys genic)	*Glossina* spp.	Biting Diptera (unimportant)	*Man.* Rarely used; pentamidine sometimes in West Africa	*Man.* Suramin in early stage; melarsoprol in late stage
ts genic)	None	Biting Diptera	*Domestic animals.* Rarely used outside Africa; as for *Duttonella* and *Nannomonas*	*Domestic animals.* Suramin; quinapyramine sulphate
:s (intra- lar or rotal), some)	None	Coitus		
ts genic)	None	Biting Diptera		

evansi.

Leishmania spp. of medical importance.

Species complex	*Species or subspecies*	*Distribution*	*Disease caused** *Primary*	*Possible secondary*	*Reservoir host*
L. donovani	*L. donovani*	Asia, tropical Africa	Visceral leishmaniasis (kala-azar)	Post kala-azar dermal leishmanoid	None (rodents in Sudan?)
	L. infantum	Mediterranean littoral, N. Africa, Middle East, China Asiatic Russia	Infantile kala-azar (rare in adults)	None	Mainly dogs a other canids
	L. chagasi	S. America	Like *L. infantum*	None	Cats, dogs, fox
L. tropica	*L. tropica*	Mediterranean, oases of N. Africa, Middle East, USSR, India	Oriental sore (dry, urban CL)	None	Dogs?
	L. major	N. and tropical Africa, Middle East, USSR	Wet, rural CL	None	Mainly gerbils
	L. aethiopica	Ethiopia, Kenya, South Yemen?	Oriental sore	Diffuse CL (DCL)	Rock hyraxes
L. mexicana	subspecies *mexicana*, *amazonensis*, *pifanoi*	S. America	CL (not *pifanoi*) (Chiclero's ulcer, bay sore)	DCL	Forest rodents marsupials
L. braziliensis	subspecies *braziliensis*, *guyanensis*, *panamensis*	S. and Central America	CL (pian-bois with *guyanensis*)	Mucocutaneous leishmaniasis (espundia) – mainly *braziliensis*	Forest rodents (also monkeys and sloths with *panamensis*
	L. peruviana	Andes (W. Peru)	CL (uta) in highlands up to 3000 m	None	Dogs

**Note*: All forms treated usually with antimonials or aromatic diamidines; response of different species an subspecies varies.
CL = cutaneous leishmaniasis

Characteristics of species of *Plasmodium* which infect man.

Species	Geographical distribution	*Duration of schizogony* Exoerythrocytic	Erythrocytic	*No. of merozoites per schizont* Exoerythrocytic	Erythrocytic	*Main morphological characteristics of erythrocytic forms* 'Ring' form	Trophozoite	Schizont	Gametocytes	Host erythrocyte[1]
P. vivax	Worldwide, in tropical, subtropical and warmer temperate regions	8 days	48 hours	~10000	12–24	At least one-third diameter of erythrocyte	Amoeboid	10 µm diameter	Round or ovoid, male: 9 µm, female: 10–11 µm.	Enlarged, stippled ('Schüffners' dots')
P. malariae	Worldwide, but scattered, mainly tropical and subtropical	14–15 days	72 hours	~15000	6–12	At least one-third diameter of erythrocyte	Compact, often bandlike	7 µm diameter	Round or ovoid, 7 µm	Not enlarged; very faint 'Ziemann's stippling' after prolonged staining only
P. ovale	Tropical Africa; also occasionally in other parts of tropics and subtropics (possibly of exogenous origin)	9 days	48 hours	~15000	6–12 (12–24 in relapses)	At least one-third diameter erythrocyte	Compact	7 µm (larger in relapses)	Round or ovoid, 9 µm	Slightly enlarged, stippled ('Schüffner's dots'); may be distorted and elongated
P. falciparum	Worldwide, in tropics, subtropics and warmer temperate regions	5½ days	48 hours	~30000	8–24	Very small at first; 2 nuclei commonly; some apparently on edge of cell (accolé forms)	Compact, rarely seen peripheral blood	5 µm; rarely seen in peripheral blood	Crescentic, male: 9–11 µm long, female: 12–14 µm	Not enlarged; often with fewer larger dots ('Maurer's clefts')

[1] These alterations do not develop until some growth of the parasite has occurred.

Note: Chemoprophylaxis with proguanil, chlorproguanil, amodiaquine, pyrimethamine or chloroquine. Chemotherapy with chloroquine, amodiaquine, quinine or pyrimethamine plus a sulphonamide. For destruction of persisting exoerythrocytic forms of *P. vivax*, primaquine is necessary.

Piroplasmia of veterinary importance.

Species	Size[1]	Geographical distribution	Main hosts: Vertebrate	Main hosts: Invertebrate	Pathogenicity	Common name of disease (if any)
Babesia bigemina	Large	Central and S. America, Europe, Africa, Australia	Cattle, deer	*Boophilus, Haemaphysalis, Rhipicephalus*	High	Red-water fever
B. bovis	Small	Europe, Russia, Africa	Cattle, deer	*Ixodes, Boophilus, Rhipicephalus*	Moderate	Red-water fever
B. divergens	Small	W. and Central Europe (including England)	Cattle	*Ixodes*	Moderate	Red-water fever
B. argentina	Small	S. and Central America, Australia	Cattle	*Boophilus*	High	
B. major	Large	Europe (inc. England?) Russia	Cattle	*Boophilus*	Low	
B. caballi	Large	S. Europe, Asia, Russia, Africa	Equidae (domestic)	*Dermacentor, Hyalomma, Rhipicephalus*	Moderate	
B. equi	Small	S. Europe, Asia, Russia, Africa, S. America	Equidae (domestic and zebra)	*Dermacentor, Hyalomma, Rhipicephalus*	High	Biliary fever
B. motasi	Large	S. Europe, Russia, Africa, Far East, tropical America	Sheep, goats	*Rhipicephalus, Haemaphysalis, Dermacentor*	Moderate	
B. ovis	Small	As for *B. motasi*	Sheep, goats	*Rhipicephalus, Ixodes*	Low	

Piroplasmia of veterinary importance, continued.

Species	*Size*[1]	*Geographical distribution*	*Main hosts* *Vertebrate*	*Invertebrate*	*Pathogenicity*	*Common name of disease (if any)*
B. trautmanni	Large	S. Europe, Africa, Russia	Pigs	*Rhipicephalus*	Moderate	
B. canis	Large	N., Central, S. America, S. Europe, Russia, Africa, Asia	Dogs and wild canids	*Rhipicephalus, Dermacentor, Haemaphysalis*	Moderate-high	Tick fever
B. gibsoni	Small	India, Ceylon, China	Dogs and wild canids	*Rhipicephalus, Haemaphysalis*	High (in domestic dog)	Tick fever
B. felis	Small	Africa, India	Domestic cat, lion, leopard	?*Haemaphysalis*	Moderate	
Theileria parva	Small	East Africa (eradicated from southern Africa)	Cattle, buffalo	*Rhipicephalus, Hyalomma*	High	East coast fever
T. annulata	Small	N. Africa, S. Europe, S. Russia, India, China	Cattle, water buffalo	*Hyalomma*	High	Mediterranean coast fever
T. mutans	Small	Africa, Asia, Europe (including England), Russia, Australia, N. America	Cattle	*Rhipicephalus, Boophilus*	Very low	
T. hirci	Small	N. Africa, S.E. Europe, S. Russia, Middle East	Sheep, goats	?*Rhipicephalus*	High	
T. ovis	Small	Africa, Europe, Russia, India, Middle East	Sheep, goats	*Rhipicephalus*	Very low	

[1] In this table, 'large' means about 2–5 μm long, and 'small' means about 1–2 μm.

Note: Babesiasis is treated with diminazene aceturate; there is no adequate chemotherapy for theileriasis.

Further Reading and References

General reference books

AIKAWA, M. and STERLING, C. R. (1974). *Intracellular Parasitic Protozoa*. Academic Press, New York.

BAKER, J. R. (1973). *Parasitic Protozoa*, 2nd edition. Hutchinson, London.

CHEN, T. C., (ed.) (1971). *Aspects of the Biology of Symbiosis*. University Park Press, Baltimore.

COX, F. E. G., (ed.) (1982). *Modern Parasitology*. Blackwell Scientific Publications, Oxford.

KREIER, J. P. (ed.) (1977–78). *Parasitic Protozoa*, 4 volumes. Academic Press, New York.

LEVINE, N. D. (1973). *Protozoan Parasites of Domestic Animals and of Man*. 2nd edition. Burgess Publishing Company, Minneapolis.

WHITFIELD, P. J. (1979). *The Biology of Parasitism*. Edward Arnold, London.

WILSON, R. A. (1979). *An Introduction to Parasitology*. Studies in Biology no. 4, 2nd edition. Edward Arnold, London.

See also Parasitology Supplement (1978). *Nature, Lond.* **273**, 596–630.

Books on various taxonomic groups

BRUCE-CHWATT, L. J. (1980). *Essential Malariology*. Heinemann Medical Books, London.

CORLISS, J. O. (1979). *The Ciliated Protozoa*, 2nd edition. Pergamon Press, Elmsford (N.Y.).

ELLIOTT, K., O'CONNOR, M. and WOLSTENHOLME, G. E. W. (eds) (1974). *Trypanosomiasis and Leishmaniasis with special reference to Chagas' disease*. Ciba Foundation Symposium 20. Associated Scientific Publishers, Amsterdam.

GARNHAM, P. C. C. (1966). *Malaria Parasites*. Blackwell Scientific Publications, Oxford.

HAMMOND, D. M. and LONG, P. L. (eds) (1973). *The Coccidia*. University Park Press, Baltimore and Butterworths, London.

HOARE, C. A. (1972). *The Trypanosomes of Mammals*. Blackwell Scientific Publications, Oxford.

LUMSDEN, W. H. R. and EVANS, D. A. (eds) (1976–79). *The Biology of the Kinetoplastida*, 2 volumes. Academic Press, London.

LYONS, K. M. (1978). *The Biology of Helminth Parasites*. Studies in Biology no. 102. Edward Arnold, London.

PHILLIPS, R. F. (1982). *Malaria*. Studies in Biology. Edward Arnold, London (in preparation).

Additional references, mainly related to various sections

1 Introduction

COLEMAN, G. S. (1980). Rumen ciliate protozoa. *Adv. Parasitol.*, **18**, 121–73.

HONIGBERG, B. M. (1970). Protozoa associated with termites and their role in digestion. In *Biology of Termites*, K. Krishna and F. M. Weesner (eds), vol. 2, pp. 1–36. Academic Press, New York.

2 *Finding and infecting the host*

CANNING, E. U. and WRIGHT, C. A. (eds) (1972). *Behavioural Aspects of Parasite Transmission*. Academic Press (for the Linnaean Society), London.

KENNEDY, C. R. (ed.) (1976). *Ecological Aspects of Parasitology*. North Holland Publishing Company, Amsterdam.

3 *Development in the host*

CHAPPELL, L. H. (1980). *Physiology of Parasites*. Blackie, Glasgow.

GARNHAM, P. C. C. (1977). The continuing mystery of relapses in malaria. *Protozool. Abstracts*, **1**, 1–12.

GUTTERIDGE, W. E. and COOMBS, G. H. (1977). *Biochemistry of Parasitic Protozoa*. Macmillan, London.

JENNI, L., MOLYNEUX, D. H., LIVESEY, J. L. and GALUN, R. (1980). Feeding behaviour of tsetse flies infected with salivarian trypanosomes. *Nature, Lond.*, **283**, 383–5.

LEWANDOWSKY, M. and HUTNER, S. H. (1979–81) *Biochemistry and Physiology of Protozoa*, 2nd edition. 4 vols. Academic Press, New York.

SCHOLTYSECK, E. (1979). *Fine Structure of Parasitic Protozoa*. Springer, Berlin.

VON BRAND, T. (1973). *Biochemistry of Parasites*, 2nd edition. Academic Press. New York.

4 *Outwitting the host's defences*

BROWN, K. N. (1976). Specificity in host-parasite interaction. In *Receptors and Recognition*, P. Cuatrecas and M. F. Greaves, (eds), pp. 121–75. Chapman and Hall, London.

COHEN, S. and MITCHELL, G. H. (1978). Prospects for immunisation against malaria. *Current Topics in Microbiol. and Immunol.*, **80**, 97–137.

COHEN, S. and SADUN, E. H. (eds) (1976). *Immunology of Parasitic Infections*. Blackwell Scientific Publications, Oxford.

COX, F. E. G. (1978). Specific and nonspecific immunisation against parasitic infections. *Nature, Lond.*, **273**, 623–6.

INCHLEY, C. J. (1981), *Immunobiology*. Studies in Biology no. 128. Edward Arnold, London.

OGILVIE, B. M. and WILSON, R. J. M. (1976). Evasion of the immune response by parasites. *Brit. Med. Bull.*, **32**, 177–81.

PLAYFAIR, J. H. L. (1978). Effective and ineffective immune responses: evidence from experimental models. *Current Topics in Microbiol. and Immunol.*, **80**. 37–64.

PORTER, R. and KNIGHT, J. (eds) (1974). *Parasites in the Immunized Host: Mechanisms of Survival*. Ciba Foundation Symposium 25. Associated Scientific Publishers, Amsterdam.

SOLTYS, M. A. (1973). A review of studies on immunization against protozoan diseases of animals. *Z. Tropenmed. Parasitol.*, **24**, 309–22.

TAYLOR, A. E. R. (ed) (1968). *Immunity to Parasites*. Symp. Brit. Soc. Parasitol., **6**. Blackwell Scientific Publications, Oxford.

TERRY, R. J. and SMITHERS, S. R. (1975). Evasion of the immune response by parasites. *Symp. Soc. Exp. Biol.*, **29**, 453–65.

VICKERMAN, K. (1978). Antigenic variation in trypanosomes. *Nature, Lond.*, **273**, 613–7.

WILLIAMS, R. O., YOUNG, J. R. and MAJIWA, P. A. O. (1979). Genomic rearrangements correlated with antigenic variation in *Trypanosoma brucei*. *Nature, Lond.*, **282**, 847–9.

WOO, P. T. K. (1974). A review of studies on the immunization against the pathogenic protozoan diseases of man. *Acta Trop.*, **31**, 1–27.

5 Damaging the host

CLARK, I. A. (1978). Does endotoxin cause both the disease and parasite death in acute malaria and babesiosis? *Lancet*, 8 July 1978, pp. 75–7.

LASER, H., KEMP, P., MILLER, N., LANDER, D. and KLEIN, R. (1975). Malaria, quinine and red cell lysis. *Parasitology*, **71**, 167–81.

LOCKEY, M. W. (1978). Primary amoebic meningoencephalitis. *Laryngoscope*, **88**, 484–503.

LOSOS, G. and CHOUINARD, A. (eds) (1979). *Pathogenicity of Trypanosomes*. International Development Research Centre, Ottawa. (Publication number IDRC–132e.)

MANSFIELD, J. M. (1978). Immunobiology of African trypanosomiasis. *Cell. Immunol.*, **39**, 204–10.

TAYLOR, A. E. R. (ed.) (1966). *The Pathology of Parasitic Diseases*. Symp. Brit. Soc. Parasitol., **4**. Blackwell Scientific Publications, Oxford.

TAYLOR, A. E. R. and MULLER, R. (eds) (1975). *Pathogenic Processes in Parasitic Infections*. Symp. Brit. Soc. Parasitol., **13**. Blackwell Scientific Publications, Oxford.

6 Leaving the host

WORMS, M. J. (1972). Circadian and seasonal rhythms in blood parasites. In *Behavioural Aspects of Parasite transmission*, E. U. Canning and C. A. Wright (eds), pp. 53–66. Academic Press (for the Linnaean Society), London.

7 Evolution

BAKER, J. R. (1974). The evolutionary origin and speciation of the genus *Trypanosoma*. *Symp. Soc. Gen. Microbiol.*, **24**, 343–66.

TAYLOR, A. E. R. (ed.) (1965). *Evolution of Parasites*. Symp. Brit. Soc. Parasitol., **3**. Blackwell Scientific Publications, Oxford.

Subject Index